Environmental Epidemiology and Risk Assessment

Tim E. Aldrich
Jack Griffith

Edited by
Christopher Cooke

VNR VAN NOSTRAND REINHOLD
———————— New York

To:
My teachers, Reuel A. Stallones, M.D.,
and Herman F. Lehman, DDS

TA

Jackie. You're the best!

JG

Copyright © 1993 by Van Nostrand Reinhold

I(T)P™ Van Nostrand Reinhold is a division of International Thomson Publishing, Inc.
The ITP logo is a trademark under license

Printed in the United States of America

For more information, contact:

Van Nostrand Reinhold
115 Fifth Avenue
New York, NY 10003

Chapman & Hall GmbH
Pappelallee 3
69469 Weinheim
Germany

Chapman & Hall
2-6 Boundary Row
London
SE1 8HN
United Kingdom

International Thomson Publishing Asia
221 Henderson Road #05-10
Henderson Building
Singapore 0315

Thomas Nelson Australia
102 Dodds Street
South Melbourne, 3205
Victoria, Australia

International Thomson Publishing Japan
Hirakawacho Kyowa Building, 3F
2-2-1 Hirakawacho
Chiyoda-ku, 102 Tokyo
Japan

Nelson Canada
1120 Birchmount Road
Scarborough, Ontario
Canada M1K 5G4

International Thomson Editores
Campos Eliseos 385, Piso 7
Col. Polanco
11560 Mexico D.F. Mexico

All rights reserved. No part of this work covered by the copyright hereon may be reproduced or used in any form or by any means—graphic, electronic, or mechanical, including photocopying, recording, taping, or information storage and retrieval systems—without the written permission of the publisher.

96 97 98 99 BBR 10 9 8 7 6 5 4 3 2

Library of Congress Cataloging-in-Publication Data

Aldrich, Tim.
    Environmental epidemiology and risk assessment / by Tim Aldrich and Jack Griffith and Christopher Cooke.
      p. cm.
    Includes bibliographical references and index.
    ISBN 0-442-00885-6
    1. Environmental health. 2. Environmentally induced diseases-
-Epidemiology. 3. Environmental monitoring. I. Griffith, Jack D.
II. Cooke, Christopher, 1944- III. Title
    (DNLM. 1. Environmental Monitoring—methods. 2. Epidemiologic methods. 3. Risk Factors. QT 140 A365e)
RA566.A43   1992
616.9'6—dc20
DNLM/DLC
for Library of Congress

92-49832
CIP

Contents

Contributors vii

Foreword ix

Preface xiii

1 Public Awareness, Federal Policy, and Environmental Epidemiology 1

OBJECTIVES 1
PUBLIC AWARENESS 1
FEDERAL POLICY 6
SUMMARY 7
ASSIGNMENTS 10

2 Epidemiology: The Environmental Influence 13

OBJECTIVES 13
THE ENVIRONMENTAL INFLUENCE
  ON EPIDEMIOLOGY 13
EPIDEMIOLOGY 18
ENVIRONMENTAL EPIDEMIOLOGY 22
SUMMARY 23
ASSIGNMENTS 24
GLOSSARY 24

3 Epidemiologic Research Methods 27

OBJECTIVES 27
INTRODUCTION 27

iv   Contents

        BASIC STUDY DESIGNS   27
        DESCRIPTIVE EPIDEMIOLOGY   29
        ANALYTIC EPIDEMIOLOGY   32
        ESTABLISHING RISKS IN ENVIRONMENTAL
            EPIDEMIOLOGY   40
        MEASUREMENT ERROR   43
        MORBIDITY AND MORTALITY MEASUREMENT   46
        STATISTICAL CONSIDERATIONS   48
        DETERMINING CAUSAL INFERENCES   54
        SUMMARY   56
        ASSIGNMENTS   56
        GLOSSARY   57
        RECOMMENDED READING   58

4 Disease Clusters   61

        OBJECTIVES   61
        BASIC STUDY DESIGNS   61
        KEY QUESTIONS WHEN CONSIDERING A
            CLUSTER REPORT   63
        THE CLUSTER SOFTWARE   66
        OTHER METHODS   73
        STRATEGIES FOR USING STATISTICS TO TEST FOR DIS-
            EASE CLUSTERING   74
        SUMMARY   76
        ASSIGNMENTS   77
        GLOSSARY   78

5 Surveillance Activities in Disease and Exposure Situations   83

        OBJECTIVES   83
        BACKGROUND   83
        DEVELOPING A SURVEILLANCE PROGRAM   84
        ANALYTIC ISSUES   91
        SENTINEL EVENTS   98
        MODIFICATIONS OF THE SURVEILLANCE
            RESEARCH DESIGN   100
        SUMMARY   101
        ASSIGNMENTS   102

6   Characterizing Human Exposure   105

OBJECTIVES   105
CHARACTERIZING HUMAN EXPOSURE   105
MAGNITUDE OF THE POTENTIAL PROBLEM   107
SOURCES OF ENVIRONMENTAL
  CONTAMINATION   108
ENVIRONMENTAL AND BIOLOGICAL MONITORING OF
  EXPOSURE   113
ABSORPTION   116
SUMMARY   125
ASSIGNMENTS   125
GLOSSARY   126

7   Laboratory Practice   132

OBJECTIVE   132
INTRODUCTION   132
LABORATORY PRACTICE   132
HUMAN TISSUES AND BODY FLUIDS   135
LABORATORY QUALITY CONTROL   141
SPECIFICITY AND SENSITIVITY OF
  LABORATORY ASSAY METHODS   142
SUMMARY   149

8   Biomarkers in Environmental Epidemiology   152

OBJECTIVES   152
INTRODUCTION   152
DEFINITIONS, CLASSIFICATION OF MARKERS,
  AND CONCEPTS   154
SURVEY OF BIOMARKERS   157
TYPE 1 BIOMARKERS   158
APPLICATION OF BIOMARKERS TO
  EXPOSURE ESTIMATION   168
APPLICATION TO PUBLIC HEALTH   170
METHODOLOGICAL CONSIDERATIONS   171
SUMMARY   175
ASSIGNMENTS   177

9   Disease and the Environment   182

OBJECTIVES   182
PUBLIC PERCEPTION   182

CARCINOGENESIS 186
REPRODUCTIVE AND DEVELOPMENTAL
  EFFECTS 195
NEUROLOGIC EFFECTS 200
SUMMARY 207
ASSIGNMENTS 208

10 Risk Assessment 212

OBJECTIVES 212
PUBLIC POLICY 212
THE RISK ASSESSMENT PROCESS 219
WORKING EXAMPLE OF THE RISK
  ASSESSMENT MODEL 230
HOW TO USE A RISK ASSESSMENT 234
SUMMARY 236
ASSIGNMENTS 237
RECOMMENDED READING 237

11 Public Communication, Participation, Risk Management 240

OBJECTIVES 240
PUBLIC COMMUNICATION 240
COMMUNITY ACTION AND INTERACTION
  WITH HEALTH AGENCIES 242
THE JOINT AGENCY AND COMMUNITY
  EPIDEMIOLOGIC STUDY 244
THE NORTH CAROLINA RULE 245
THE PUBLIC HEALTH FUNCTION 248
THE RISK MANAGEMENT FUNCTION 251
SUMMARY 254
ASSIGNMENTS 254
SUGGESTED READINGS 255

12 Legal Aspects of Environmental Epidemiology 259

ASSIGNMENTS 264

Index 267

Contributors

Tim E. Aldrich, Ph.D., M.P.H.
North Carolina Department of Environment,
Health, and Natural Resources
Raleigh, North Carolina

Jack Griffith, Ph.D. M.S.
Raleigh, North Carolina

Christopher Cooke, M.S.
School of Public Health
University of North Carolina at Chapel Hill
Chapel Hill, North Carolina

Wanzer Drane, Ph.D.
Department of Epidemiology and Biostatistics
University of South Carolina
Columbia, South Carolina

Vincent F. Garry, M.D.
School of Medicine
University of Minnesota
Minneapolis, Minnesota

Robert C. Duncan
School of Medicine
University of Miami
Miami, Florida

Charles H. Nauman, Ph.D.
Environmental Monitoring Systems Laboratory
U.S. Environmental Protection Agency
Las Vegas, Nevada

Jerry N. Blancato, Ph.D.
Environmental Monitoring Systems Laboratory
U.S. Environmental Protection Agency
Las Vegas, Nevada

Robert Meyer, Ph.D.
N.C. Dept. Env. Health and Nat. Res.
Raleigh, North Carolina

Robert Gustafson, Ph.D.
Department of Religion and Philosophy
Pembroke University
Pembroke, North Carolina

David Graber, M.P.H.
Department of Health Planning and Policy
University of North Carolina at Chapel Hill
Chapel Hill, North Carolina

Darlene Meservy, Dr.PH., R.N.
Department of Family Medicine
University of Utah
Salt Lake City, Utah

Jay A. Meservy, J.D.
Salt Lake City, Utah

Foreword

John R. Goldsmith, M.D., M.P.H.

The scientific approach to human problems has had an erratic history, interacting as it has with superstition and magic, conventional wisdom, politics, economic power, and religion.

Among these problems, environmental challenges have been prominent. Much of the history of human culture is bound up in the quest for sufficient food, adequate shelter and clothing, and protection from threats of other species. Imbedded in this culture are the results of a sort of folk epidemiology, the experience of human communities with risk factors for disease and injury. The nineteenth and twentieth centuries have witnessed not so much of a change, as an intensification of these processes. As the science on which society's defensive strategies could be based have had increasingly wide applications, technology has been responsible for an increasing fraction of the environmental risks that are prominent today. Some of these risks such as depletion of the ozone layer and the "greenhouse effect of atmospheric warming" are now global rather than local challenges. Even a technological disaster such as Chernobyl has potential impacts on Europe and much of the Northern Hemisphere.

Public attention and anxiety, often fueled by radio, television, and the press, not only may exaggerate the risk, and thus distort the allocation of protective resources, but they can also catalyze a better scientific assessment. An interesting example is the discovery of excess leukemia among residents in Seascale, United Kingdom, first brought to scientific and public attention by the British Independent Television. This led to the recognition that the excess could be accounted for by children of fathers who worked in and had increased radiation exposures at the Sellafield Nuclear Center. The recognition of the probable perinatal time of exposure and the excess risk for children of exposed workers represented an exemplary epidemiological contribution of Gardner (1991) and his associates.

At the same time, much of the current practice of radiological or environmental health protection is based on the systematic development of risk assessment. Baseline risk assessment in a recent monograph (U.S. EPA 1989) on toxic waste sites is said to include **data collection and evaluation** for a specified site, **exposure assessment, toxicity assessment, and risk characterization.** Risk assessment is usually based on extrapolation from toxicological information. The manual refers to the possible value of epidemiological studies in a deprecatory fashion: "For most Superfund sites, studies of human exposure or health effects in the surrounding population will not be available. However, if controlled epidemiological or other health studies have been conducted, perhaps as a consequence of the preliminary ATSDR health assessment of other community involvement, it is important to include this information in the baseline risk assessment as appropriate. However, not all such studies provide meaningful information in the context of Superfund risk assessment."

Thus, the discipline of environmental epidemiology does not have much status among risk assessors, whose confidence in extrapolation from experimental findings seems limitless. However, Bross (1990), in a provocatively titled article, has pointed out an important scientific problem in the use of dose-response information for health protection. He notes that agents causing mutations, such as radiation can have two different effects on cells, depending on whether the cell survives. If it survives with a mutation, its progeny may develop uncontrolled growth, which in the organism is manifest as cancer. However, if the cell is killed, no effects of its progeny are possible. From this he derives the principle that if exposures of a cell to a mutagen can have two counterbalancing effects, the dose-response relationship will have a maximum. Under these circumstances, with increasing dose the response, say cancer, may first increase and then decrease. In fact, many examples of such reactions occur. To the extent that they do, linear extrapolation of high dose-response relationships to low doses will substantially underestimate the possible effects.

Over such issues as this tension can be expected.

The authors are veterans of such tensions and have prepared in this book a guide to how to respond constructively to them. Without books such as these, such issues are at risk of being either overlooked or overwhelmed by emotion and invective.

Other sources of tension exist within the discipline of environmental epidemiology. They include the dilemma of how to respond to low increased risks that may be highly statistically significant, but have low attributable risks, compared to other risks that may have higher risk ratios and attributable risks but low significance. Examples abound of studies said to have "negative" results, because a predetermined arbitrary level of significance of 0.05 was not reached, for which the power of a study to have a significant finding was insufficient. A final source of potential tension can be found in the attitude toward "positive" findings. The growing interest in using epidemiology for monitoring (Goldsmith 1984) implies that many monitoring studies will produce no evidence for unfavorable effects. To the extent that such results may indicate the soundness of protective measures, the findings have a positive social impact, even though they may not advance science very far.

Tensions between disciplines, between weak and strong effects, and between protective and investigative uses of environmental epidemiology are going to be with us for a long time. This book is a source of information and insight in dealing with these tensions. We have needed it and are grateful for the authors' patience in providing it in such a well-organized fashion.

REFERENCES

Bross, I. 1990. "How to Eradicate Fraudulent Statistical Methods: Statisticians Must do Science." *Biometrics* 46:1213–1225.

Gardner, M.J. 1991. "Father's Occupational Exposure to Radiation and the Raised Level of Childhood Leukemia Near the Sellafield Nuclear Plant." *Environmental Health Perspect.* 94:5–7.

Goldsmith, J.R. 1984. Editor "Epidemiological Monitoring in Protection from Environmental Health Hazards" a Special Issue of Science of the Total Environment 32:211–363.

U.S. Environmental Protection Agency. 1989. *Risk Assessment Guidance for Superfund: Volume I, Human Health Evaluation Manual* (Part A), Interim Final. EPA/540/1-89/002.

Preface

The impetus for this book came from our students, specifically at the University of Utah, who were concerned with learning to apply epidemiologic methods to environmental exposures, and from our own experience as epidemiologists working with such exposures. As John Goldsmith so eloquently said in the Foreward, environmental challenges are constantly with us, and more are appearing daily. We believe that the practice of environmental epidemiology may well be the tool that enables public health professionals to meet these demanding challenges.

Although this book can serve as a reference text for trained epidemiologists, it was developed for an additional purpose. Clearly, this book is not for theorists. We have placed considerably more emphasis than is commonly found in epidemiologic texts on exposure monitoring, the use of biomarkers in epidemiologic studies, risk assessment, and risk management. Understanding full well the importance that public opinion plays in the success or failure of any important regulatory decision involving human studies, we have attempted to address the issue of communicating scientific findings to the public. This **"how to"** book is also designed for public health professionals with limited didactic training in epidemiology, but with interests and professional responsibilities that require a working understanding of the discipline.

For those readers who intend to use this book as a class text, a set of case studies will be available from the publisher. These case studies are based on actual environmental "episodes," and are designed to supplement discussion and didactic presentation of material in this initial text.

1

Public Awareness, Federal Policy, and Environmental Epidemiology

Tim E. Aldrich and Jack Griffith

OBJECTIVES

This chapter will:

1. Discuss how public awareness of environmental issues has developed in the United States.
2. Describe how federal environmental policy has been a basis for the development of environmental epidemiology as a risk assessment tool.

PUBLIC AWARENESS

Public awareness of pollution-related health problems has grown over recent years. This increased awareness has led to public concern that continued damage to the environment is dangerous and that steps to prevent its occurrence should be implemented. There is also an emerging conviction that environmentally associated health problems should be identified, measured, and remedial activities undertaken. This awakening concern on the part of the public is due, in no small measure, to the reporting of environmental "incidents" by the popular media. We have briefly described five such environmental episodes in the following pages, including the public's response, as examples of how media coverage influences popular perception of health risk.

The Love Canal, New York (1979)

From 1947 to 1952, the Hooker Chemical company buried an estimated 21,800 tons of industrial refuse in an abandoned canal in upstate New York. In 1953, Hooker sold the Love Canal site to the Niagara School Board for one dollar (accompanied by a deed that disclaimed any responsibility for any injuries that might result from the buried wastes). The school board built an elementary school on the site and sold the remaining land for residential development. Around 1979, chemicals from the dump site began to leach into surrounding homes and the school, and complaints of unusually high frequencies of miscarriages, birth defects, and cancer began to surface.

As a result of public concern, several health studies were conducted in the neighborhood. Vianna and Polan (1984) reported increased incidence of low birth weight babies among Love Canal residents who were characterized as exposed based on their house being in an area that was highly susceptible to ground water percolating to the surface. Goldman et al. (1985) found an excess of low birth weight babies born in the Love Canal area when compared to controls. In a 1985 paper, Paigen et al. reported that the children living in Love Canal homes, when compared with controls, experienced significantly more seizures, learning problems, hyperactivity, eye irritation, skin rashes, abdominal pain, and incontinence. Paigen and Goldman (1987), however, found no significant difference in prematurity among Love Canal residents who were characterized as exposed based on the distance of their homes from the dump site and on the proximity of the home to potential chemical migration. These investigators did report an increased incidence of birth defects among the exposed populations. Neighborhood concern over cancer was reasonable considering that more than 200 chemicals (NRC 1991a), including the carcinogens benzene and γ-hexachlorobenzene (NRC 1991b) were found to be stored in the dump site. However, an ecologic study (Janerich et al., 1981) found no significant increase in the incidence rates of any specific cancer, including liver, lymphoma, or leukemia.

Although low birth weight was the primary chronic adverse health effect reported in excess for Love Canal residents, as more information concerning the pervasiveness of toxicants in the canal was made public, health concerns of the neighborhood residents became so acute that President Jimmy Carter declared a state of emergency in the area, clearing the way for the relocation of 710 Love Canal families (Kolata 1980).

Three Mile Island, Pennsylvania (1979)

A rupture in the containing wall of a reactor vessel at a nuclear power plant released low-level radioactive gas into the ambient environment. The plant was closed while repairs were made to the reactor. Monitoring of the environment around the reactor, and in the ambient environment outside the plant, indicated that any release of radiation had been insignificant with regard to potential adverse health effects. Follow-up analyses also confirmed that there were absolutely no adverse health consequences related to the released radioactivity.

There remains, however, considerable public concern over the operation of nuclear reactors because of the potential for a meltdown (for example, the reactor overheats, the shell containing the nuclear radiation cracks precipitating an explosion, and deadly radiation is released into the ambient environment).

Chernobyl, Soviet Union (1986)

The public's concern regarding nuclear power plants was heightened following the experience in the (former) Soviet Union. In 1986, a nuclear reactor in Chernobyl overheated, melting the radioactive core. Radiation released into the ambient environment made a wasteland of the area surrounding the plant, and literally spread throughout the world as radioactive dust reached high into the atmosphere. Deaths among workers and nearby residents occurred within hours, and related deaths (for example, cancer related to the exposure) are taking place today. Although widespread cleanup followed, the area surrounding Chernobyl remains highly contaminated.

Immediately following the accident, there was an attempt on the part of the Soviet government to deny its occurrence. However, international monitoring activities soon forced the Soviet government to acknowledge the disaster. Soviet authorities then attempted to minimize the problem. Despite the official attempts at damage control, media attention focused on the human suffering related to this terrible environmental disaster and brought it to the attention of people all over the world.

Seveso, Italy (1976)

Chlorophenoxy herbicides, including 2,4,5-T and 2,4-D, have been used throughout the United States for more than 30 years to control weeds, broadleaf plants, and trees on farms, livestock ranges, public rights-of-way, parks, recreation areas, and forests (*Federal Register* 1978). During the 2,4,5-T manufacturing process, TCDD (2,3,7,8-tetrachlorodibenzopara-dioxin), is produced. TCDD, also known as dioxin, is an extremely toxic chemical (LD 50 reported to be between 0.6 and 115 µg/kg for several animal species) (Menzer and Nelson, 1980).

Because of its chemical persistence and lipophilic nature, it is possible that

TCDD could enter the human food chain. For example, Kaczmar et al. (1983) found TCDD residues ranging from 17 to 586 ng/kg in fish taken from several Michigan rivers. Gross (1983) found residues in the part per trillion (ppt) level in human breast milk among Oregon women and in the tissue of a human conceptus.

At the time of the Seveso incident, about all that was known about TCDD came from experimental studies. Dioxin had been associated with teratogenic effects in laboratory animals (Moore and Courtney 1971), increased rates of spontaneous abortions in rhesus monkeys (McNulty 1977), and increased tumors in rodents (Van Miller et al. 1977). The only confirmed human health effect was a chronic skin eruption (chloracne) associated with long-term occupational exposure (Poland et al. 1971), although there were some data to suggest that workers exposed to dioxin were at increased risk to soft-tissue sarcoma.

On July 10, 1976, a safety dish on a reaction vessel in a 2,4,5-T manufacturing plant in Seveso ruptured, releasing 3 to 16 kg of TCDD into the atmosphere (it was reported that the plume reached 30 to 50 m high over the factory, and then settled on an area about 2 km long and 700 m wide). As a result of contact with this chemical, people began to feel ill very soon after exposure, followed within days, by chloracne, stomach pains, and internal bleeding.

Soon after the explosion, surrounding communities were evacuated in a series of concentric circles, each representing levels of suspected contamination. The soil of the innermost zone was excavated and removed. At the time, due to the chemical persistence of TCDD, it was felt that it would take as long as ten years before the area could be resettled. A study by Bissanti et al. (1983) of the Seveso population reported that spontaneous abortions experienced a sharp increase early in 1977. Subsequently, Fara and Del Corno (1985) reported similar findings, with pregnancy loss rates among the Seveso population peaking during the first quarter of 1977, and leveling off thereafter to 1976 levels.

Times Beach, Missouri (1979)

In 1979, residents of Times Beach Missouri began to notice unexplained and frightening things were happening in their community. Horses were getting sick or having skin eruptions. Other animals were dying, and some were born with severe birth defects. Birds were even reported to be falling out of the sky. Investigators subsequently determined that waste oil, purchased from a chemical recycling plant and containing high levels of TCDD, had been deposited on several rural roads and on the dirt in a horse stable as a means of settling dust.

Health studies conducted in the area shortly after recognition of the problem by the Missouri State Health Department did not show any exposure-related health effects in the community. However, the combined response of the residents, environmentalists, and news media resulted in a climate of fear that was reported around the country. An overwhelming public outcry led regulatory authorities to purchase the property, relocate the residents, and dredge the soil to clean it of the toxic wastes at a cost of millions of dollars.

About this time, TCDD was also identified as a component in Agent Orange (a deforestation product used extensively in Viet Nam). Agent Orange became the subject of a class action law suit by soldiers who served in that country and believed that subsequent illnesses (for example, cancer, birth defects) were due to exposure during their tour of duty.

Woburn, Massachusetts (1986)

An epidemiologic study (Lagakos, et al.) in 1986 reported an association between childhood leukemia and exposure to drinking water from two wells contaminated by a toxic waste disposal site in Woburn, Massachusetts. Originally, 12 cases of leukemia were diagnosed with 5.3 cases expected. The cases were identified between 1964 (the start-up date for pumping the wells) and 1983. The wells were closed in 1979. In a subsequent study in 1986, an additional 8 cases were identified; the total cases (20) were statistically in excess of the 9.1 cases expected from national rates.

The toxic waste site included a large pit containing animal hides, solvents, and other chemical wastes. A nearby, abandoned lagoon, was heavily contaminated by lead, arsenic, and other metals. Trichloroethylene, tetrachloroethylene, and chloroform were found in the well water. Residues of trichlorethylene, a known carcinogen widely used as an industrial solvent in degreasing and extraction processes as well as in the dry cleaning industry, were also found in drinking-water wells. Levels of trichlorethylene were 30 to 89 times higher than the recommended EPA Maximum Contamination Level (MCL) of 5 parts per billion (ppb). Although the agent was not experimentally associated with the type of cancer in question, it has been found to cause hepatocellular carcinoma in mice. A highly sensationalized trial followed the release of the data from this study, and there were several large financial awards for the families of all the cited cancer cases.

Several important points can be made from reviewing these episodes:

1. The role of the media is prominent in all the episodes. The media brought the episode to the public's attention and then provided a forum for an exchange of information prior to remediation.

2. With regard to Love Canal and Times Beach residents, public perception is a significant consideration. The EPA was consistently reminded of the public concern over potential risks to health by the news media and special interests groups.
3. In the Seveso and Times Beach incidents, sentinel events (health effects among animals) were the first signs of potential human risks.
4. The public's perception of adverse health outcomes associated with involuntary exposure to dangerous substances in the environment is frequently formed on the basis of a few well-publicized episodes (Harris 1984).
5. Media coverage of these unfortunate, and sometimes tragic environmental episodes, has contributed to a crisis in confidence regarding the public's perception of governmental responsiveness to adverse environmental exposures (Allman 1985).

As can be seen from these episodes, environmental health problems are frequently called to the attention of the government by concerned and frightened citizens, often in a blaze of media visibility. Unfortunately, in this type of environment, health professionals, risk assessors, and risk managers find it increasingly difficult to communicate to the public. As a result, health risks believed to be associated with a specific environmental exposure may not be clearly understood by the public.

FEDERAL POLICY

In the past three decades, public concern about protecting the environment has grown dramatically and has become a potent force in the enactment of legislation designed to protect the environment. In response to environmental awareness on the part of the public, the Congress of the United States has enacted legislation with profound influence on the genesis of environmental epidemiology. The Comprehensive Environmental Recovery, Conservation, and Liability Act (Superfund) is specifically designed to evaluate and control the environmental impact of hazardous waste sites. Some of the federal statutes enacted into law since 1969, and designed to protect the environment, are shown in Table 1-1. Under the Superfund Amendments and Reauthorization Act of 1986 (SARA), the Agency for Toxic Substances and Disease Registry (ATSDR) was implemented to enforce environmental regulations and conduct research related to environmental pollutants. Federal regulations often determine the extent and availability of study populations and of data sources for research activities. Federal agencies are also responsible for determining the environmental impact of chemical pollution, and this evaluation includes protection of public health. In fact, a majority of scientists who investigate environmen-

TABLE 1-1. **Environmental Legislation Since 1969**

Law	Date Enacted
Environmental Protection Act	1969
Occupational Health and Safety Act	1970
National Cancer Institute Act	1971
Consumer Product Safety Act	1972
Federal Water Pollution Control Act	1972
Clean Air Act	1976
Federal Insecticide, Fungicide, and Rodenticide Act	1976
Resource Conservation and Recovery Act	1976
Toxic Substances Control Act	1976
Comprehensive Environmental Recovery, Conservation, and Liability Act	1980
Superfund Amendments and Reauthorization Act	1986.

tally related problems are supported through a variety of federal programs designed to protect the environment.

A fair question to ask might be: "Is all this governmental regulation in response to a real problem?" The easy answer to this question is "Yes." Clearly, we can't keep polluting our environment without suffering severe consequences in the coming generations. The great difficulty with this answer is that it dictates remedial action (for example, finding a way to assess risks without unduly alarming the population or angering the industries that provide the economic livelihood for much of the country's population). Environmental epidemiologic studies that are properly designed and executed can do much to establish an understanding of environmental risk assessment. Through the years (see Chapter 2) epidemiology has been the principal means of identifying the various causes of human disease. The discipline continues to be an indispensable component in detecting, validating, quantifying, and monitoring environmental disease determinants.

SUMMARY

Over the past decade, there has been a decline in the quality of the environment and a loss of respect on the part of the public regarding the capacity, and intent, of governmental authorities to control polluters. The public is frequently presented with complex horror stories, purported to be scientifically reliable, concerning the risks associated with a certain environmental exposure. Such stories are often encapsulated into a brief paragraph in the newspaper or a 30 second sound bite on television or radio (Shodell 1985), and are designed to hold attention as much as to inform. Highly visible law suits associated with hazardous occupational exposures to selected chemicals have also led to a generalized anxiety when the same agents occur in the ambient environment (for example, asbestos, polyvinyl chloride, polychlorinated biphenyls [PCBs], and dioxin).

This tendency in the media to sensationalize each discovery of a new agent that may cause cancer has promoted the coining of two provocative phrases: "cancerphobia" and "carcinogen of the week." Both of these abstractions convey the public's perception of cancer as the leading health concern from environmental contamination, and the notion that the ambient environment is a "sea" of carcinogens. Such a repercussion to each new scientific advance reported in the mass media has made many scientists and agency representatives reticent in communicating their findings.

Public health and regulatory agencies are regularly accused of doing a poor job in identifying and communicating environmentally related health problems to the public in a responsible and understandable fashion. They often respond by citing the complexity of the problems associated with identifying and managing environmental adverse health effects (for example, the myriad risk factors, the intricate disease processes, and the complex nature of host–environment interactions). Although the response may be entirely accurate and truthful, the perception of the public is often that the agency is waffling on a potentially critical issue. Similarly, when an agency is required by law to perform a benefit analysis as part of the risk appraisal process (to see if benefits of exposure outweigh risk), it is considered to be pro-industry.

What general truths can we gather from this discussion? There are clearly difficulties in dealing with complicated scientific issues in a public forum. However, the public has every right to expect that agencies charged with protecting its health be responsive and communicative. At the heart of the problem for public health professionals is the lack of specialized training for conducting studies of these potential environmental health risks (Ozonoff and Boden 1987; Bender et al. 1990). In this respect, disease prevention has expanded to add a new level (Figure 1-1): prevention of disease from ambient exposures (Nasseri 1979).

In the classic "levels of prevention" paradigm, tertiary prevention is after disease has occurred and has impacted on the individual, the objective of prevention is to prevent unnecessary suffering and untimely death. In secondary prevention, the disease has occurred, but has not impacted the person. The objective of secondary prevention is early detection and appropriate treatment so that a person is restored to normal health. Primary prevention is directed before the disease has occurred. In this new paradigm (Figure 1-1), this level is subdivided into interventions directed to personal risk factors (for example, diet and lifestyle)—Category

II and prevention actions directed to (ambient) environmental hazards—Category I.

Unfortunately, we still do not begin to understand all there is to know about estimating health risk from environmental exposures. Even with the development of important laboratory methods for identifying and measuring biological markers of toxic exposure and effect, research methods for exposure assessment are often inadequate or inappropriately applied in epidemiologic studies.

To complicate matters further, environmental epidemiology is often conducted in a crisis environment amid highly politicized and emotionally charged settings. There is often disagreement among the public, scientists, and certainly among responsible authorities as to the definition of a health risk. For this reason, regulatory agencies are involved in developing new approaches to assessing health risks (for example, gathering and analyzing data on hazard identification, dose–response estimation, exposure assessment, and risk characterization). More about risk assessment in Chapter 10.

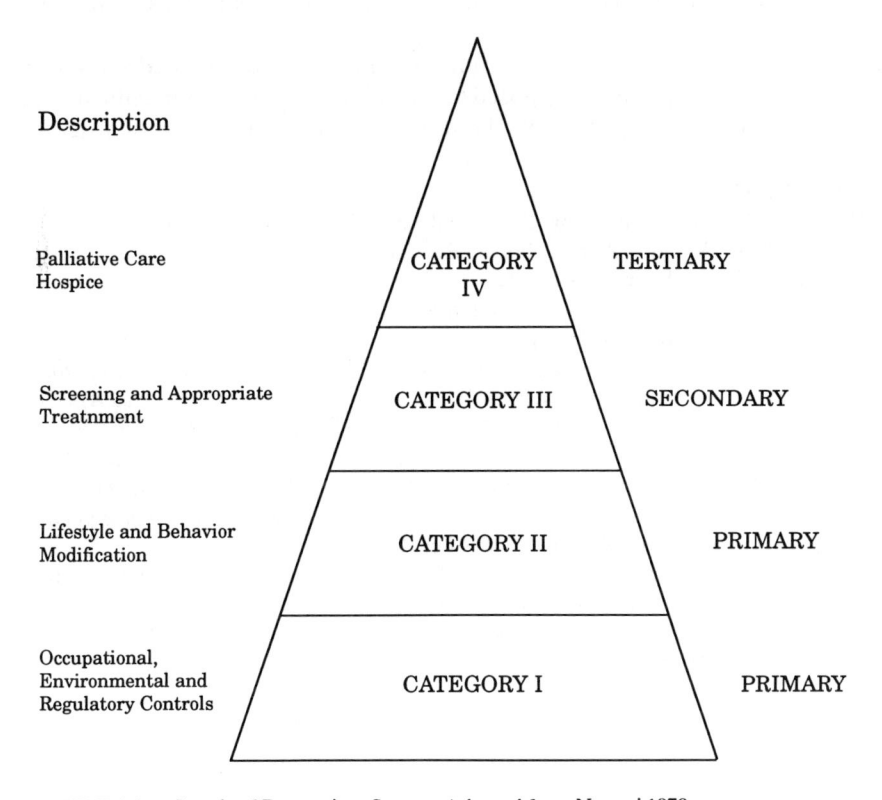

Description		
Palliative Care / Hospice	CATEGORY IV	TERTIARY
Screening and Appropriate Treatment	CATEGORY III	SECONDARY
Lifestyle and Behavior Modification	CATEGORY II	PRIMARY
Occupational, Environmental and Regulatory Controls	CATEGORY I	PRIMARY

FIGURE 1-1. Levels of Prevention. Source: Adapted from Nasseri 1979.

10   Environmental Epidemiology

Good data in the risk assessment process will provide a better understanding of the process of estimating the potential health impact of environmental pollution (Chapter 11). Clearly, a well-designed and conducted epidemiologic study provides the best available means of gathering human data for risk assessment purposes, and the best predictors of risk remain those that are based on aggregated epidemiologic observations (Squire 1981).

ASSIGNMENTS

1. Review the legislative acts in Table 1-1. Consider the election year effect, since many of these bills were passed in presidential election years.
2. Consider the "carcinogen of the week" perception that exists in the general populace on the basis of fears for environmental contamination. Evaluate this "external" source of risk in light of the differences between Category I (ambient environment) and Category II (personal environment) presented in Figure 1-1. Address (in a one-page narrative) this "phobia" on the basis of the individual's ability to control his or her own risks.
3. Look in a newspaper or magazine for an example of the media's role in informing the public, as opposed to sensationalizing environmental health concerns. Write a paragraph about how you react to this sort of article.

REFERENCES

Allman W. F. "We have Nothing to Fear (but a Few Zillion Things)." *Science* 85, 38–41.

Bender, A. P., Williams, A. N., Johnson, R. A., and Jagger, H. G. 1990. "Appropriate Public Health Responses to Clusters: The Art of Being Responsibly Responsible." *American Journal of Epidemiology* (132):S48–S52.

Bissanti, L., Pignatti, C. B., Marni, E., Abate, L., Basso, P., Formigaro, F., Strigini, P., and Santi, L. 1983. *Final Report on the Congenital Defects and Other Unfavourable Outcomes of Pregnancy Found in the Population of the Seveso Area Affected by TCDD Pollution on 10/7/76.* Malformation Registry, Department of Seveso.

Fara, G. M., and Del Corno, G. 1985. "Pregnancy Outcome in the Seveso Area After TCDD Contamination. Prevention of Physical and Mental Congenital Defects." In: *Part B: Epidemiology, Early Detection and Therapy, and Environmental Factors.* ed. Marois, M. New York: Alan R. Liss, Inc.

Federal Register [6560-01] Vol. 43, No. 78-Friday, April 21, 1978. Environmental Protection Agency, Office of Pesticide Programs, Washington, D.C.

Gross, M. 1983. "Ultratrace analyses of Tetrachlorodibenzodioxin (TCDD) in Environmental Samples." In: *Final Report for Cooperative Agreement No. CR-806847.* University of Nebraska, Lincoln, and the U.S. EPA.

Harris, D. 1984. "Health Department: Enemy or Champion of the People." *American Journal of Public Health* 74: 428–430.

Heinlein R. 1983. *Job.* New York, Berkley Press.

Janerich, D. T., Burnett, W. S., Feck, G., Hoff, M., Nasca, P., Polednak, A. P.,

Greenwald, P., Vianna, N. 1981. "Cancer Incidence in the Love Canal Area." *Science* 212:1404–1407.
Kaczmar, S. W., Zabik, M. J., and D'Itri, F. M. 1983. *Part per Trillion Residues of 2,3,7,8-Tetrachlordibenzo-para-dioxin in Michigan Fish.* Presented to the Division of Environmental Chemistry, American Chemical Society.
Kolata, G. B. 1980. "Love Canal: False Alarm Caused by Botched Study." *Science* 208:1239–1242.
Lagakos, S. W., Wessen, B. J., Zelen, M. 1986. "An Analysis of Contaminated Well Water and Health Effects in Woburn, Masscahusetts." *J. Am. Stats. Assoc.* 81:583–596.
McNulty, W. P. 1977. "Toxicity of 2,3,7,8-Tetrachlorodibenzo-p-dioxin for Rhesus monkeys: Brief Report." *Bull. Environ. Contam. Toxicol.* 18:108–109.
Menzer, R. E., and Nelson, J. O. 1980. "Water and Soil Pollutants." In *Casarett and Doull's Toxicology: The Basic Science of Poisons.* 2nd Ed. ed. Doull, J., Klaassen, C. D., and Amdur, M. O. New York: Macmillan Publishing Co., Inc.
Moore, J. A., and Courtney, K. D. 1971. "Teratology Studies With the Trichlorophenoxyacid Herbicides, 2,4,5-T and Silvex." *Teratology* 4:236.
Nasseri K. 1979. (Letter to the Editor), *International Journal of Epidemiology* 8(4);389–90.
National Research Council. 1983. Commission on Life Sciences. Committee on the Institutional Means for Assessment of Risks to Public Health. *Risks Assessment in the Federal Government: Managing the Process.* Washington, DC: National Academy of Sciences.
National Research Council. 1991a. *Environmental Epidemiology: Volume I, Public Health and Hazardous Wastes.* Washington, DC: National Academy of Sciences, p. 138.
National Research Council. 1991b. *Environmental Epidemiology: Volume I, Public Health and Hazardous Wastes.* Washington, DC: National Academy of Sciences, p. 184.
Ozonoff, D. and Boden, L. 1987. "Truth and Consequences: Health Agency Responses to Environmental Health Problems." *Science, Technology, & Human Values* 1987 Summer/Fall, 70–77.
Paigen B., Goldman L. R. 1987. "Lessons from Love Canal, New York, USA. The Role of the Public and the Use of Birth Weights, Growth, and Indigenous Wildlife to Evaluate Health Risk." In *Health Effects from Hazardous Waste Sites*, JB Andelman and DW Underhill, eds. Chelsea, Michigan: Lewis, pp. 177–192.
Paigen, B., Goldman, L. R., Highland, J. H., Magnant, M. M., Steegman, A. T. 1985. "Prevalence of Health Problems in Children Living Near Love Canal. *Haz. Wastes Haz Materials* 2:23–43.
Poland, A. P. Smith, D., Metter, G., and Possick, P. 1971. "A Health Survey of Workers in a 2,4-D and a 2,4,5-T Plant With Special Attention to Chloracne, Porphyria Cutanea Tarda, and Psychologic Parameters." *Arch. Environ. Health* 22:316–327.
Shodell, M. 1985. Risky Business. *Science* 85:43–47.
Squire, R. A. 1981. "Ranking Animal Carcinogenesis: A Proposed Regulatory Approach." *Science* 214:877–80.

U.S. Environmental Protection Agency (EPA). 1989. *A Management Review of the Superfund Program.* Washington, D.C.: U.S. Environmental Protection Agency.

Van Miller, J.P., Lalich, J.J., and Allen, J.R. 1977. "Increased Incidence of Neoplasms in Rats Exposed to Low Levels of 2,3,7,8-Tetrachlorodibenzo-p-dioxin." *Chemosphere* 6:537–544.

Vianna, N. J., and Polan, A. K. 1984. "Incidence of Low Birth Weight Among Love Canal Residents." *Science* 226:1217–1219.

2
Epidemiology: The Environmental Influence

Jack Griffith and Tim E. Aldrich

OBJECTIVES

This chapter will:

1. Provide a historical overview of epidemiology
2. Define environmental epidemiology, and
3. Discuss the environmental influence on epidemiology.

THE ENVIRONMENTAL INFLUENCE ON EPIDEMIOLOGY

Prior to the development of the "germ theory," the science of bacteriology, and the identification of bacteria as causative factors in human disease, man believed that his physical environment was often responsible for illness. Even early civilizations noted an association between seasonal changes in climate, temperature, moisture, overcrowding and filth, and the onset of disease. Unfortunately, resulting illnesses in these times were often characterized by the rapid onset of symptoms, and all too frequently, death.

> "Whoever wishes to investigate medicine properly should proceed thus: in the first place to consider the seasons of the year, and what effects each of them produces. Then the winds, the hot and the cold, especially such as are common to all countries, and then such as are peculiar to each locality. In the same manner, when one comes into a city to which he is a stranger, he should consider its situation, how it lies as to the winds and the rising or setting of the sun. One should consider most attentively the waters which the inhabitants use, whether they be

marshy or soft, or hard and running from elevated and rocky situations, and then if saltish and unfit for cooking; and the ground, whether it be naked and deficient in water, or wooded and well watered, and whether it lies in a hollow, confined situation, or is elevated and cold; and the mode in which the inhabitants live, and what are their pursuits, whether they are fond of drinking and eating to excess, and given to indolence, or are fond of exercise and labor."

Hippocrates penned these words in 427 B.C. in his treatise *On Airs, Waters, and Places*, and they encompass the very essence of environmental epidemiology (Krieger 1972). Hippocrates understood, intuitively perhaps, that the environment plays an important role in relation to human health. Further, it appears that he recognized that illness could be identified and treated, not only within the patient, but within the patient's environment. In his practice, Hippocrates observed and counted disease occurrence in populations, and attempted to associate disease with environmental factors such as work and home environments, water, and climate. From Hippocrates's writings, the doctrine of miasma and vapors developed, and survived until the end of the nineteenth century. The proponents of this theory believed that disease was caused by mysterious poisonous substances that rose up from the earth and were spread through the winds. For example, observers of the time noted that persons living near swamps were more likely to die from fever. They believed that the fever occurred as a result of exposure to "bad" air emanating from the swamps, and the disease was subsequently known as malaria: bad (mal) air (aria). For his foresight, Hippocrates is considered by many to be the first great epidemiologist.

Throughout our history, we have been plagued by diseases for which we have had no understanding of either the cause or the means of transmission. In primitive times, since there was no obvious cause for an illness, it was only natural to subscribe it to the work of an evil spirit, or an angry god. If a man was seen to act like a maniac, he was thought to be possessed of the devil. God was also thought to punish sinners through disease and pestilence. In I Chron. 21:1–14, when David displeased Jehovah by numbering the children of Israel, there was an epidemic of pestilence and in three days, "there fell of Israel 70,000 men." In the middle ages when plague, smallpox, and cholera killed millions of people, it was believed that God was punishing humanity for its sins. As recently as colonial times, smallpox was believed to be the vengeful act of God.

Although religious and miasmatic theories of disease were prominent until the end of the nineteenth century, people began to notice a person-to-person relationship with the spread of disease. In medieval times, travel was sometimes restricted because travelers were known to be coming from areas where disease was thought to be prevalent. During the sixteenth century, persons ill with the plague, smallpox, or yellow fever were isolated and simply left to die because of the fear that they might somehow spread the disease to others. And, of course, everyone knows about the segregation of lepers, a fear-motivated behavior that remains real, even today.

As early as 1546, an Italian physician and poet named Hieronymus Fracastorius, in *De Res Contagiosa,* hypothesized that infection was transferred from person-to-person by minute, invisible particles. In 1662 John Graunt, in his treatise *Natural and Political Observations Mentioned in a Following Index and Made upon the Bills of Mortality,* was the first person to compile data on births, deaths, and illnesses in London (Table 2-1). Graunt's was the first known attempt to use morbidity and mortality data as a basis for vital statistics.

During the same period in London, Thomas Sydenham, an English physician who is primarily known for his use of an opium derivative (laudanum) as a pain killer, investigated the occurrences of various epidemics in relation to the season of the year and age of the victim. In 1762, a prominent Vienna physician named Anton von Plenciz published a treatise in which he suggested that living organisms were responsible for specific diseases in humans. In 1799, Noah Webster, a lawyer, published *Epidemic and Pestilential Diseases* in which he attributed the great epidemics of yellow fever, scarlet fever, and influenza to a combination of environmental factors affecting large numbers of people. Jacob Henle, a German scientist, theorized in 1840 that tiny microbes or germs caused disease. Henle's supposition became known as the germ theory.

In 1837 William Farr, a physician in London, used available statistical data to test various epidemiologic theories of his time, including the miasmatic cause of cholera. But the seminal event in the emergence of environmental epidemiology was the investigation of cholera toward the end of the nineteenth century by an English physician named John Snow. In his description of cholera, Snow suggested that the disease "...invariably commences with the affection of the alimentary canal...there can be no doubt that these symptoms depend upon the exudation from the mucous membrane, which is soon afterwards copiously evacuated...into the sewage system, and finally into the drinking water supplies of many citizens." Snow hypothesized that a living cell, smaller than the eye can see and living in the exudate, entered the public water supply by "...permeating the ground, and getting into wells, or by running along channels and sewers into the rivers from which entire towns are sometimes supplied with water (Snow 1855)."

Snow used mortality data supplied by William Farr to count, measure, and analyze the distribution of cholera in London (Table 2-2). He observed that eight to nine times more cholera deaths occurred in areas of London that were supplied with water by the Southwark and Vauxhall Company.

TABLE 2-1. **An excerpt from John Graunt's** *Natural and Political Observations Mentioned in a Following Index, and Made Upon the Bills of Mortality.*

The Diseases, and Casualties This Year Being 1632

Abortive and Stillborn	445	Jaundies	43
Afrighted	1	Jawfain	8
Aged	628	Impostume	74
Ague	43	Kil'd by Accident	46
Apoplex, and Meagrom	17	King's Evil	38
Bit with a mad dog	1	Lethargie	2
Bleeding	3	Livergrown	87
Blood flux, Scowring, & Flux	348	Lunatique	5
Bruse; Issues, Sores, Ulcers	28	Made away themselves	15
Burnt and Scalded	5	Measles	80
Burst, and Rupture	9	Murthered	7
Cancer, and Wolf	10	Over-laid, and starved at nurse	7
Canker	1	Palsie	25
Childbed	171	Piles	1
Chrisomes, and Infants	2,268	Plague	8
Cold, and Cough	55	Planet	13
Colick, Stone, and Strangury	56	Pleurisie, and Spleen	36
Consumption	1,797	Purples, and Spotted Fever	38
Convulsion	241	Quinsie	7
Cut of the Stone	5	Rising of the Lights	98
Dead in the street, and starved	6	Sciatic	1
Dropsie and Swelling	267	Scurvey, and Itch	9
Drowned	34	Suddenly	62
Executed, and Prest to death	18	Surfet	86
Falling Sickness	7	Swine Pox	6
Fever	1,108	Teeth	470
Fistula	13	Thrush, and Sore Mouth	40
Flox, and Small Pox	531	Tympany	13
French Pox	12	Tissick	34
Gangrene	5	Vomiting	1
Gowt	4	Worms	27
Grief	11		

Males 4,494	Males 4,932	Whereof, of
Christened Females 4,590	Buried Females 4,603	the Plague—8
In All 9,584	In All 9,535	

Source: Adapted from Fox, Hall, and Elveback 1970.

He further noted fewer deaths in areas of London supplied by the Lambeth Company. Snow subsequently determined that the Lambeth Company took its water from a section of the Thames that was far above London, and was, as he put it "...quite free from the sewage of London." During one "terrible episode," Snow observed that a large number of people dying from cholera lived near a community water pump on Broad Street, an area of London

TABLE 2-2. Deaths in the Population Supplied by the Southwark and Vauxhall Water Company and and the Lambeth Water Company, London 1853.

	No. of Homes	Deaths From Cholera	Deaths in Each 10,000 Homes
Vauxhall Co.	40,046	1263	315
Lambeth Co.	26,107	98	37
Rest of London	256,423	1422	59

Source: Adapted from Goldsmith 1986.

that received most of its water supply from the Southwark and Vauxhall Company. Putting his theory into practice, Snow persuaded city officials to remove the handle of the Broad Street pump, and the epidemic subsided (Figure 2-1). Although he performed a significant contribution to the public health of the citizens of London by showing that cholera in London was associated with the source of drinking water, it was through the establishment of a causal chain linking environmental contamination to the spread of disease that Snow made his lasting contribution to the emergence of environmental epidemiology (Mausner and Bahn 1974).

In the late nineteenth century the compound microscope was developed, and during this period of scientific enlightenment Joseph Lister, Louis

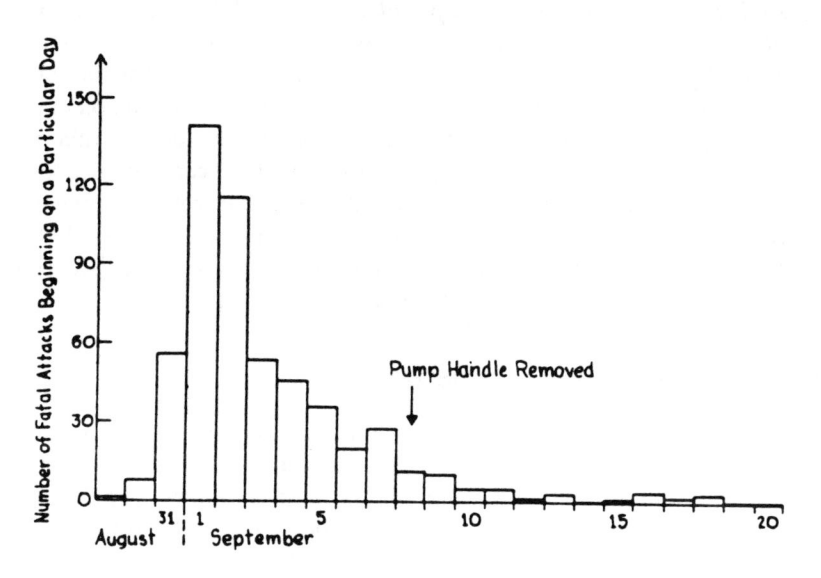

FIGURE 2-1. Snow's Epidemic Curve. Source: Goldsmith 1986.

Pasteur, and Robert Koch ushered in the science of bacteriology (the study of the nature and behavior of bacteria). With this constantly expanding knowledge came a greater understanding of the causative nature of diseases. In 1882, Robert Koch, in his *Postulates* hypothesized that before a germ could be said to cause a disease, the following factors would be required:

1. The germ must always be found in every case of the disease.
2. The germ is not found with any other disease.
3. The germ must be isolated from someone with the disease, cultured through several generations in the laboratory, and produce the disease under experimental conditions.

These *Postulates* have become the basis for understanding environmental agents that are now considered to be the classic causes of disease. However clever Koch's hypothesis, in the early and middle 1800s no one could test them since microscopes of the period were imperfect, and laboratory methods to study microbes were not available.

EPIDEMIOLOGY

Epidemiology has been variously defined as *the science of the occurrence of disease* (Anderson et al. 1962) and *the study of the distribution of a disease or a physiological condition in human populations and of the factors that influence this distribution* (Lilienfeld 1978). We would suggest the following definition: **Epidemiology is the study of the occurrence and distribution of disease and injury specified by person, place, and time.** The word epidemiology is derived from two Greek words: epi (upon) and demos (people). Epidemiology can be defined literally to be that which descends upon people, and originally was restricted to the study of diseases characterized by sudden occurrence in epidemic fashion (an abnormal number of cases). Subsequently, however, the definition has broadened to include the existence of disease in any form and its prevalence in the population.

Evolution of Epidemiologic Reasoning

At its basic level, epidemiology as a discipline is used to quantify risk (for example, the likelihood of an event occurring as a result of another factor). Some refer to this as cause and effect, others as determinism (a person does or does not develop disease). However, it appears that determinism is too concise for processes involving human populations; simple dichotomies operate at the level of the individual, not in a population (Kleinbaum et al.

1982). Epidemiologists use probabilistic data to establish causal relationships in populations (Susser 1973). Fundamental to this process is the proposition that risk is uniform (Zeger 1991). Using analytic tenets of person, place, and time, epidemiologists sift through disease patterns to determine characteristics associated with high risk. Once the high risk characteristics are determined, they may be modified, eliminated, or at least identified with the goal of preventing further disease occurrence.

Posing the "right question" is contrast to epidemiologic reasoning. Hypotheses for assessing characteristics associated with disease risk may be developed using Mills's *Canons* (Roht et al. 1982):

1. Specific Agreement: What do the cases have in common to the exclusion of the noncases?
2. Specific Difference: What do the noncases have in common to the exclusion of the cases?
3. Concomitant Variation: Does the disease occurrence vary in accord with another factor (for example, the seasons, trends)?
4. Analogy: Does this pattern remind you of something you have seen before?

Using these and similar deductive processes, epidemiologists endeavor to evaluate characteristics of individual cases as well as compare disease rates between population groups to obtain estimates of risk (Armenian 1991).

Infectious Disease Epidemiology

Early epidemiologic practice focused on infectious (or communicable) diseases. The term infection implies that an organism capable of causing a disease is present and multiplying in the body. Disease implies that a viable interaction has taken place between the organism and the infected body. For an infectious disease to occur, there must be (1) a reservoir of the specific etiologic agent, (2) the agent must escape from the reservoir and find its way to the new host, and (3) the host must be susceptible. Through the years a number of infectious diseases have plagued man, and have been effectively characterized by the respective causative organism: Animal parasites (for example, protozoa, metazoa that are small single cell, or multicell animals such as Giardia), and bacteria (for example, single cell organisms that are biologically intermediate between animals and plants, such as Escherichia coli) that are found in water supplies with fecal contamination. These agents, upon transmission to a susceptible host, cause severe intestinal disorders, such as diarrhea.

Viruses, representing an ultramicroscopic life form, are known to cause measles, mumps, smallpox, rabies, and poliomyelitis. Rickettsiae are viruslike organisms that cause typhus fever and Rocky Mountain spotted fever. Plant parasites such as molds and fungi are also responsible for causing epidermophytosis (athlete's foot) and coccidioidomycosis. Tuberculosis (TB), a destructive disease of the lungs, is due to an infection by the tubercle bacillus (Mycobacterium tuberculosis). Tuberculosis is spread from infected to uninfected persons. The body responds to the infection of the bacterium by producing fibrous tissue or, in some cases, calcification that replaces the necrotic tissue. A person's progress depends on the rapidity with which the diseased tissue is replaced.

During the last part of the eighteenth century, almost all the bodies that were autopsied showed some evidence of active or healed TB. This was the time in which TB was not only the leading cause of death, but little was known about its transmission. Networks of hospitals and clinics dedicated to the treatment of TB were spread throughout the United States. With the advent of effective diagnostic and therapeutic techniques, the epidemic dissipated. Today, TB is found in less than 10 percent of young adults in this country, and is no longer considered to be a viable threat to most communities within the United States. However, certain groups of inner city residents, and persons living in developing countries, remain at great risk to this disease. For example, a new and apparently very virulent strain of TB seems to be appearing in urban areas of this country in persons with active acquired immune deficiency syndrome (AIDS). Thus far, this new strain of TB is fatal in a majority of cases, and is seemingly resistant to the standard forms of prophylactic treatment.

Poliomyelitis (polio), is an acute systemic infection caused by a neurotropic virus which attacks the central nervous system (CNS), and frequently causes a variable degree of permanent paralysis. The virus is fairly resistant to chemicals and will live for several weeks in an appropriate media (for example, sewage). The reservoir of infections is in all cases the infected person. Eleven distinct and severe outbreaks of polio were recorded in the United States between 1907 and 1949. In 1916, there were more than 27,000 cases recorded in the United States and more than 7,000 deaths. Today, thanks to the development of the Salk vaccine in the mid-1950s (an effective immunization against polio), the disease is confined to sporadic outbreaks involving very small numbers of people in this country. However, populations in developing countries are at greater risks where appropriate preventive measures are not practiced.

Eventually these diseases, along with other acute infections, succumbed to conventional public health measures of quarantine and vaccination. Other infectious diseases that were subdued by primary prevention efforts are:

1. Influenza, a respiratory disease, characterized by fever, malaise, and marked prostration, sometimes with inflammation of the upper respiratory tract, caused by a specific virus (A, B, and Asian strains).
2. Measles, a viral infection characterized by fever, rash, and respiratory symptoms.
3. Diphtheria, an illness characterized by local infection of the tonsils or larynx and a general toxemia, caused by the bacillus Corynebacterium diphtheria.
4. Pertussis (whooping cough) an infection of the respiratory tract, characterized by severe coughing and gastrointestinal distress, caused by the bacillus Hemophilus pertussis.
5. Smallpox (variola) a viral disease that is characterized by fever, malaise, and a generalized eruption on the layers of the skin that progresses through the stage of papules to vesicles and pustules.
6. Rheumatic fever, an acute and frequently recurring infection of unknown etiology, characterized by fever and swelling of the joints. Although rheumatic fever is, in and of itself rarely fatal, its sequelae (for example, severe damage to the heart) causes this disease to remain a major infectious disease problem in childhood and early adolescent life.

During the period of infectious disease study, although microbiology was in its inception, organisms were being cultured and vaccines developed. During this time as well, genetic studies were beginning, and laboratory practices were improved (for example, the use of biological assays began during this time). Interestingly, a newly prominent infectious disease has arisen out of the classical communicable disease mode AIDS. Acquired immune deficiency syndrome is a highly lethal, sexually transmitted infectious disease (with a long latency period) that attacks the immune system of its host. Hopefully, this virulent organism may yet lend itself to prevention through vaccination.

Chronic Disease Epidemiology

As a result of the virtual control of infectious disease, toward the middle of this century epidemiologists began to study chronic diseases. Because of the long latency period between onset and manifest disease, the study of chronic diseases was not readily amenable to conventional epidemiologic methods. However, over the years coronary artery disease and cancer have become major areas of epidemiologic research. In 1971, the protracted suffering associated with a group of diseases, collectively identified as cancer, and the public's fear of that disease, influenced the president of the United States, Richard M. Nixon, to declare a "War on Cancer." Since that time, millions of dollars have been directed toward the cause and the cure of this deadly group of diseases.

Studies of disease processes flourished during this era. Biologic assays came into acceptance with the emergence of prominent animal cancer models to study causation (for example, tobacco smoke residue and abraded rat skin) and treatment procedures (for example, the inbred strains of mice with spontaneous tumors). Molecular technology exploded as toxicologic and pharmacokinetics mechanisms were developed. Other chronic diseases, such as cerebrovascular disease, and the neurologic diseases (for example, the muscular neuropathies) also emerged during this period as important areas of epidemiologic research. With the emergence of the chronic disease phase of epidemiology came an awareness that risk factors associated with an individual's life style (for example, the type and amount of food we eat; stress related to occupational and home environments; what and how much we drink; whether we smoke, and how much; occupational and home exposures to toxic chemicals) may play an important role in the etiology of various chronic diseases.

Smoking tobacco has also been identified as a risk factor for cancer of the lung. Smoking has also been linked to increased risk of cerebrovascular disease, and bladder and cervical cancer. Coffee drinking has been linked to pancreatic cancer and saccharin (an artificial sweetener) to bladder cancer. Elevated blood pressure, high serum cholesterol levels, and a diet heavy in fatty acids have been identified as risk factors for coronary artery disease, and for cancers of the colon and rectum. Environmental and occupational exposures to toxic chemicals and radiation have also been linked to the etiology of certain cancers. Vinyl chloride, for example, has been linked to hepatic angiosarcoma; asbestos to malignant mesothelioma; benzene to leukemia; arsenic to respiratory cancer; trihalomethanes to bladder cancer; electromagnetic fields to leukemia; radiation to breast cancer; radiation and radiation by-products (radon) to lung cancer; and TCDD to soft-tissue sarcoma.

Unfortunately, we have also discovered that some highly sensitive people may have a genetic predisposition or susceptibility to a specific environmental insult. (For example, the nevoid basal cell carcinoma syndrome, which predisposes to radiation-induced skin cancer). In this group of people, the development of basal cell carcinoma is accentuated, occurring only months after the most innocuous X-ray therapy. Similiar genetic susceptibility is suspected for some chemicals due to individual variation in metabolic pathways or enzyme systems (for example, aromatic hydrocarbons, some pesticides).

ENVIRONMENTAL EPIDEMIOLOGY

Environmental epidemiology may be defined as **the study of environmental factors that influence the distribution and determinants of disease in human**

populations. There are two basic approaches to the study of environmental epidemiology. One approach involves the identification of a disease or injury, the characterization of an exposed and control population by selected attributes (age, sex, race, socioeconomic status, place of employment), and an attempt to establish a causal chain by reasoning from effect back to exposure. The other approach involves the identification of an exposure, the usual characterization of the population by exposed and controls, and the search for a likely health endpoint. In this scenario, exposure rather than disease drives the study.

This attempt to characterize exposure, without a well-defined disease, is an important alteration in the traditional, disease-driven epidemiologic approach. Unfortunately for environmental epidemiologists, characterizing exposure is difficult because humans live and work in many environments, and are exposed to complex mixtures of toxic pollutants at home, at work, and in the ambient environment. In fact, the lack of good exposed data has been called the *Achilles heel* of environmental epidemiology (Perera and Weinstein 1982). This shift in focus, from disease to exposure, requires a fundamental rethinking of the usual approach in developing epidemiologic studies. Although study designs employed in the traditional epidemiologic approach will continue to be used, exposure assessment must be factored into study designs at the beginning, as an equal partner. To this extent it is necessary that the most effective means of identifying and quantifying exposure be used to establish exposure and dose measurements. Sometimes these measurements will be infinitesimal, as low as parts per trillion (ppt) or in relatively larger values (for example, in the part per million [PPM] range). Sophisticated monitoring activities must also be used in the ambient environment, and sensitive laboratory assays employed for residue analyses in target organ human tissues and fluids.

SUMMARY

The emergence of many of the prominent chronic diseases that we experience today are the likely result of humanity's conquest of most of the infectious diseases described in this chapter. We now live long enough to experience chronic illness as a manifestation of personal behavior or as a result of exposure to toxic pollutants. Many factors entered into the increase in chronic disease, including an overall decline in the quality of our ambient environment through legal and illegal discharge of toxic wastes into the air, soil, rivers, and estuaries. Throughout history, we have observed an environmental relationship with disease. Disease was seen to occur at certain times of the year (seasons); where people lived and moved from place to place (migration); following environmental disasters (floods, heat, and cold); with the occupation of the victim. These observations at first formed the foundation of epidemiologic thinking, and are now being employed by modern epidemiologists in the development of environmental epidemiology. In this chapter we have touched on the historical basis for of epidemiology and the environmental influence on the development of epidemiology as a discipline.

ASSIGNMENTS

1. One of the great issues surrounding environmental epidemiology is the public's perception of health risks associated with environmental hazards. In the following list from *Science 85* (Allman 1985) (Table 2-3), a comparison is made of the way scientists and nonscientists ranked the risk of dying in any year from various activities and technologies. Review this comparison and consider the reasons for some of the differences in these rankings.
2. Consider Snow's classic work with cholera. Review Table 2-2 and Figure 2-1. Write one paragraph that addresses Snow's work as an observer of events that impact human health, and another paragraph that speaks to the use of epidemiology as a mechanism for preventing disease.

GLOSSARY

Communicable: The spreading of disease as the result of transmission by an infectious agent.
Contagion: The transmission of an infectious agent by direct contact.
Infection: The multiplication of an infectious agent in intimate relationship to a host.
Vector: The means by which an infectious agent is transferred from an infected to a susceptible host.
Vital Statistics: Data that relate to birth, death, marriage, divorce, and illness.

REFERENCES

Allman, W. F. 1985. "We Have Nothing to Fear (but a Few Zillion Things)." *Science* 85:38–41.
American Cancer Society (ACS). 1991. "Cancer Statistics for 1991" *Cancer: A Cancer Journal of Clinicians* 41:1.
Anderson, G. W., Arnstein, M. G., Lester, M.R. 1962. *Communicable Disease Control.* New York: The Macmillan Co., p. 14.
Armenian, H. K. 1991. "Case Investigation in Epidemiology." *American Journal of Epidemiology* 134(10):1067–1072.
Fox, J. P., Hall, C. E., and Elveback, L. R. 1970. *Epidemiology: Man and Disease.* London: The MacMillan Co., p. 23.
Goldsmith, J., 1986. *Environmental Epidemiology.* Boca Raton Florida: CRC Press, p. 46.

Kleinbaum, D. G., Kupper, L.L., and Morgenstern, H. 1982. *Epidemiologic Research.* Belmont, CA.: Wadsworth.

Krieger. 1972. Hippocrates, "Of Airs, Waters and Places." In *The Genuine Works of Hippocrates: Translated from the Greek.* Huntington, N.Y., 19.

Lilienfeld, A. M. 1978. *Foundations of Epidemiology.* New York: Oxford University Press, p. 3.

Mausner, J. S., and Bahn, A.K. 1974. *Epidemiology: An Introductory Text.* Philadelphia: W. B. Saunders Company.

Perera, F., and Weinstein. 1982. "Molecular Epidemiology and Carcinogen-DNA

TABLE 2-3. Priority Ranking for Risk of Death by Environmental Hazards and Technologies by Scientists and the Public

Public		Experts
1	Nuclear Power	20
2	Motor vehicles	1
3	Handguns	4
4	Smoking	2
5	Motorcycles	6
6	Alcoholic beverages	3
7	General (private) aviation	12
8	Police work	17
9	Pesticides	8
10	Surgery	5
11	Fire fighting	18
12	Large construction	13
13	Hunting	23
14	Spray cans	26
15	Mountain climbing	29
16	Bicycles	15
17	Commercial aviation	16
18	Electric power (non-nuclear)	9
19	Swimming	10
20	Contraceptives	11
21	Skiing	30
22	X-rays	7
23	High school and college football	27
24	Railroads	19
25	Food preservatives	14
26	Food coloring	21
27	Power motors	28
28	Prescription antibiotics	24
29	Home appliances	22
30	Vaccinations	25

Source: Adapted from *Science 85* (Allman 1985).

Adduct Detection: New Approaches to Studies of Human Cancer Causation." *Journal of Chronic Disease* 35:581–600.

Roht, L. H., Selwyn, B. J., Holguin, A. J., Jr., and Cristensen, B. L. 1982. *Principles of Epidemiology: A Self-Teaching Guide.* New York: Academic Press.

Snow, J. 1855. *On the Mode of Communication of Cholera (2nd ed.).* Churchill, London. Reproduced in Snow on Cholera. Commonwealth Fund, New York, 1936. Reprinted by Hafner, New York, 1965.

Susser, M. 1973. *Causal Thinking in the Health Sciences: Concepts and Strategies of Epidemiology.* New York: Oxford University Press.

Zeger, S. L. 1991. Statistical Reasoning In Epidemiology. *American Journal of Epidemiology* 134(10):1062-1066.

3
Epidemiologic Research Methods

Jack Griffith, Tim E. Aldrich and Robert C. Duncan

OBJECTIVES

This chapter will:

1. Review basic study designs for epidemiologic research,
2. Discuss measurement error,
3. Present approaches for the determination of risk, and
4. Discuss the determination of causal inferences.

INTRODUCTION

In this chapter we will briefly touch on the major epidemiologic methods and analytical tools needed to conduct an environmental epidemiologic study. For a more complete discussion of these analytical and design issues we refer the reader to any number of excellent introductory text books in epidemiology (Lilienfeld and Lilienfeld 1980; Monson 1980; Kleinbaum et al. 1982; Kahn 1983; Hennekens and Buring 1987).

BASIC STUDY DESIGNS

Epidemiologic studies are designed to find associations between exposure (cause) and disease (effect), and are based on the principle that **exposure must occur before disease**. Recall that in Chapter 2 we defined Epidemiology as the study of the occurrence and distribution of disease and injury specified by **person, place and time.** With this definition as a guide, we can place environmental epidemiologic studies into two broad design categories

(Rothman 1986): **Experimental** and **Observational**. The outline by Monson (1980) should be useful in helping the reader categorize the epidemiologic methods into a workable framework.

EXPERIMENTAL
- Clinical Trials
- Field Trials
- Community Trials

OBSERVATIONAL
- Descriptive
- Analytic
    - Longitudinal Approach
        - Cohort
            - Retrospective
            - Prospective
            - Historical Prospective
        - Case-control
    - Cross-sectional Approach

It is important that the reader clearly understands the difference between these designs: in the experimental approach, the investigator controls or manipulates conditions (for example, exposure) of the study. In the observational approach, the investigator only observes what occurs. The investigator has no control over the study population. Occasionally, natural environmental events produce conditions that would have occurred in an experimental setting: this is known as the natural experiment (Snow's study is a classic example of a natural experiment).

Experimental designs involve clinical trials (using patients as subjects); field trials (using healthy subjects in a controlled environmental situation (for example, testing a new vaccine); and community intervention trials, where a new process (for example, ozonated treated drinking water) is evaluated to determine its effectiveness in relation to other treatment modes or disease outcome. In this text, however, we will focus on the observational approach since it is most often used in environmental epidemiology. The approach to observational epidemiology can be placed into two basic designs: descriptive and analytic. Descriptive epidemiology is the study of the amount and distribution (for example, what, who, how much, when and where) of a disease in a population. Analytic epidemiology is the study of the causative factors of disease in a population.

Study subjects must be able to be identified (classified) by the study factors (exposure and disease) of interest. You may ask, "Do persons with a certain exposure experience more or less disease than persons without that exposure?"; or "Do people with a specific disease have more exposure to a factor of interest than persons without the disease?" To answer these questions, epidemiologic data are usually placed in a 2 X 2 table as shown in Table 3-1 (the same four groups will always be formed). Exposure is defined as contact with any suspect etiological factor; disease connotes any biological outcome, or event.

DESCRIPTIVE EPIDEMIOLOGY

The descriptive (observational) approach is confined to the use of existing data sources (e.g., birth, death and disease registries), and the unit of measurement is the population described by person, place, and time characteristics. Too often people attempt to associate causality with descriptive epidemiology. **This is wrong**. Descriptive studies are exploratory studies that describe a population of interest vis-a-vis potential health problems. The data may be used for hypothesis generating studies—*not* for hypotheses testing studies. Repeated analyses of the same data set with rearrangements of the study variables are in keeping with the hypotheses-generating role of this design. Yet one must be very careful that these so called "fishing expeditions" don't result in purely random, **statistically significant** events.

For example, the descriptive data (person, place, and time) presented in Table 3-2 show that rates for cancer of the lung in the decade 1970-1979 for white males in New Jersey and New Hampshire are three to five times as high as those for females. While the reader may believe that the higher rates for males during this decade reflect (1) the 20 to 30 year latency period for the disease, and (2) the preponderance of male smokers during those generations, these data do not support a causal explanation for the difference in rates for example, these are group data with no information provided on potential risk factors such as smoking history).

Descriptive data, for use in more sophisticated study designs, can also be

TABLE 3-1. Classification of Epidemiologic Study Subjects

		DISEASE	
		PRESENT	ABSENT
EXPOSURE	PRESENT	a Disease Present Exposure Present	b Exposure Only
	ABSENT	c Disease Only	d Exposure Absent Disease Absent

TABLE 3-2. The States of New Jersey and New Hampshire Lung Cancer Mortality Rates by County for White Males and Females, 1970–1979.

Lung Cancer	Mortality Rate Male	Mortality Rate Female	Ratio of Male to Female Rate
Atlantic	69.5	20.6*	3.37
Bergen	64.9	16.7*	3.88
Burlington	68.7	17.4*	3.94
Camden	79.9*	17.2*	4.64
Cape May	66.6	18.1*	3.67
Cumberland	65.2	14.8	4.40
Essex	61.4	15.7	3.91
Gloucester	72.0*	17.0	4.23
Hudson	79.3*	16.5	4.80
Hunterdon	59.1	17.8	3.32
Mercer	70.9*	15.7	4.51
Middlesex	76.4*	16.7	4.57
Monmouth	67.6	19.9*	3.39
Morris	61.5	14.7	4.18
Ocean	75.5*	19.9*	3.79
Passaic	63.4	15.7	4.03
Salem	67.0	15.8	4.24
Somerset	70.6	15.6	4.52
Sussex	75.0*	15.6	4.80
Union	62.4	14.8	4.21
Warren	62.1	13.3	4.66
NJ State	68.4*	16.6*	4.12
Belknap	76.7	14.9	5.14
Carroll	61.5	14.5	4.24
Cheshire	66.7	15.3	4.35
Coos	77.5*	17.9	4.32
Grafton	66.6	14.6	4.56
Hillsborough	71.4*	15.0	4.76
Merrimack	64.4	12.6	5.11
Rockingham	65.6	15.8	4.15
Strafford	79.8*	14.8	5.39
Sullivan	59.8	16.0	3.73
NH State	69.0*	15.0	4.60

* Deaths were used to determine statistical significance because of the instability of rates (the smaller the denominator, the larger the rate). The number of deaths is statistically significantly more than expected when compared with U.S. population.
Adapted from Riggan et al. (1983).

used to classify workers, and others, who are exposed to environmental toxicants. For example, Griffith and Duncan (1983) compared data on pesticide levels in the blood and urine of a probability sample of the U.S. population, from the Health and Nutrition Examination Survey (NHANES), with pesticide levels in blood and urine taken from citrus fieldworkers exposed to a wide variety of organophosphorus compounds. In this study, a total of 597 urine samples from Florida citrus workers were analyzed for the presence of the following organophosphate metabolites: dimethyl phosphate (DMP), dimethyl thiophosphate (DMTP), dimethyl dithiophosphate (DMDTP), diethyl phosphate (DEP), diethyl thiophosphate (DETP), and diethyl dithiophosphate (DEDTP).

The study population was divided into two groups: heavily exposed (applicators) and lightly exposed (pickers). The means and standard errors of the alkyl phosphate residues among the field-workers are shown in Table 3-3. Among the field-workers, the highly exposed, as expected, generally had the highest residue values. Although there is clearly a gradient (higher values among the citrus field-workers to lower values among the HANES sample) shown in these data, they are descriptive of person, place, and time, and not causal.

TABLE 3-3. Urinary Alkyl Phosphate Residues (PPM) Among Florida Citrus Field-workers and the HANES National Sample (Non–detectable and Trace Values Excluded From Calculations of Means and Standard Errors)

	DMP	DMTP	DMDTP	DEP	DETP	DED
Number in sample	332	323	331	332	331	331
Spray percentage above trace	48.2	17.6	6.9	68.7	33.2	23.0
Mean	.16	.08	.11	.41	.37	.24
(SE)	(.035)	(.017)	(.019)	(.068)	(.079)	(.057)
Number in sample	264	262	265	265	262	262
Harvest percentage above trace	6.1	3.8	6.8	43.0	26.3	5.3
Mean	.39	.15	.25	.09	.07	.06
(SE)	(.198)	(.083)	(.106)	(.007)	(.006)	(.006)
Number in sample	596	585	596	597	593	593
Total percentage above trace	29.5	11.5	6.9	57.3	30.2	15.2
Mean	.18	.09	.17	.30	.25	.21
(SE)	(.037)	(.019)	(.049)	(.046)	(.050)	(.049)
Number in sample	6,894	6,895	6,895	6,894	6,895	6,895
HANES percentage above trace	10.2	5.3	0.3	6.2	5.2	0.1
Mean	.05	.06	.05	.04	.04	.11

Source: Griffith and Duncan (1985).

ANALYTIC EPIDEMIOLOGY

Analytic (observational) Epidemiology may be divided into two basic approaches:

1. Longitudinal (cohort and case-control); and
2. Cross-sectional

In the longitudinal approach, subjects are identified on the basis of either having an exposure, or having a disease. In the cross-sectional approach, all persons in a given population are included in the study—that is, they are selected **without regard to exposure or disease.** Cases (disease prevalence) are identified in the population, and compared with the remainder of the population (controls). Exposure data are generally collected at the time of disease ascertainment, this parallel classification of exposure and disease is the primary difference between longitudinal and cross-sectional study designs. The fundamental difference between the longitudinal and cross-sectional designs is **in the ability to accurately assess the time sequence between exposure and effect.** In the longitudinal approach it is possible to determine that exposure precedes effect, while in the cross-sectional approach, **exposure and disease are measured at the same point in time.**

The Longitudinal Approach

The Cohort Design. For epidemiologic purposes a cohort may be defined as a group of persons who share a common exposure within a defined period of time. Cohort studies are based on registers of people (existing, or concurrently developed), identified through various means (for example, people taken from licensure boards, occupational rosters, insurance rolls, tax rolls, voter registration rolls, Social Security rolls, Medicare rolls, and disease registries). The ability to calculate disease-incidence rates in study populations is an important strength of the cohort design. The temporal concepts of the cohort designs are shown in Figure 3-1).

There are two basic cohort designs:the retrospective cohort design and the prospective cohort design. With the **retrospective cohort design**, or looking backward approach, the investigator compares a group of persons with a disease and a group of persons without the disease with respect to an exposure of interest. The study cohort can be developed from workers or others with available records on the basis of **past exposure** to a suspect etiologic factor. Analysis is completed by comparing exposure rates between the disease present and control (disease absent) groups within the population (Figure 3-2). If, as is frequently the case, no control population is available, rates in the exposed population are compared to general population rates.

Epidemiologic Research Methods   33

Cohort Study Designs

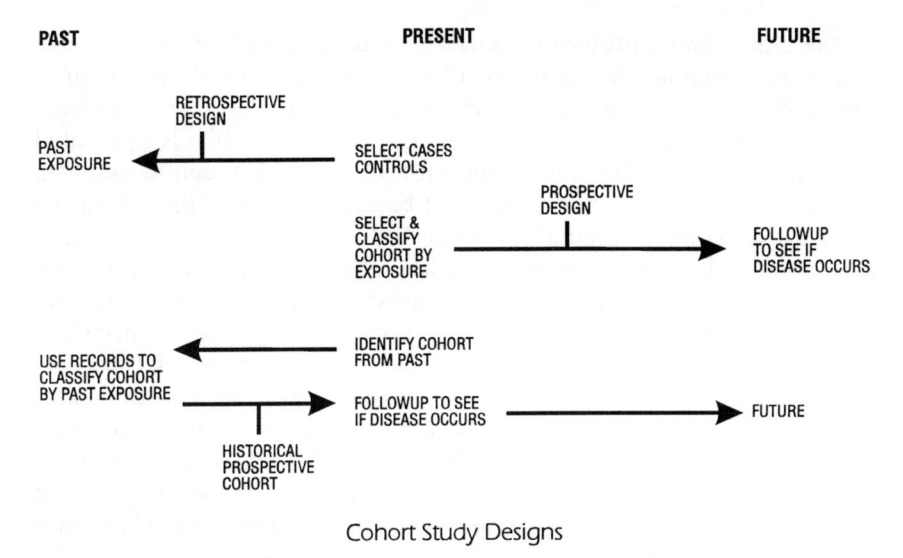

FIGURE 3-1. Cohort Study Designs.

There are several benefits to the retrospective design that should not be overlooked. Using this approach makes it possible to study many diseases in relation to one exposure (Monson 1980); the study can be completed in a much shorter time frame since the study population is being followed from a **fixed** time in the past until the present; costs are much less than the prospective design; and precise data collection, and powerful analytic procedures enhance the usefulness of this design for hypothesis testing.

		DISEASE PRESENT	DISEASE ABSENT
EXPOSURE	PRESENT	a	b
	ABSENT	c	d
		a + c	b + d

Exposure rate among disease present group $= \dfrac{a}{a+c}$

Exposure rate among disease absent group $= \dfrac{b}{b+d}$

FIGURE 3-2. Retrospective Study and Related Calculations.

There are several problems associated with the use of the retrospective design. For example, the reliability of exposure data must always be questioned. Exposure is historical, and difficult to characterize accurately from records or survivor (or surrogate) memory. Confounding bias is a potential problem in retrospective studies, since it is often very difficult to ascertain information on other factors that might be associated with the risk of the disease in question (for example, smoking history). Observation bias is a also a potential problem in retrospective studies. Observation bias occurs because the investigator is usually working with medical records containing historical information, with little likelihood of confirmation: the decision to include or exclude a medical record is based on observer interpretation, and this may vary from one observer to another. Provider data may also be highly subjective (strong emotions are often involved in recalling the circumstances of an illness or a death) (Baker et al. 1988; Howe 1988; Schechter et al. 1989).

A common modification to retrospective studies is the use of more than one comparison subject for each case subject. One comparison subject may be selected to represent the general population (random); another to be similar in age, race, etcetera; and still another for a comparable life style (for example, a neighbor). Each comparison group provides a different type of exposure that the cases may possess which are associated with the disease of interest.

The prospective cohort design is somewhat analogous to the experimental approach since the investigator is involved in the selection of a concurrent study population to follow forward in time. There is, however, one major difference between the experimental and the prospective approach: Although the study groups are selected on the basis of presumed exposure to the etiologic factor, in a prospective study the investigator has no control over which persons are assigned to exposed and unexposed groups. Persons are simply characterized as having or not having an exposure, and then are followed forward over time until disease occurs (length of follow-up will vary depending on the rarity of the disease in the population: the rarer the disease, the longer the follow-up). As participants are followed forward, cases are identified, and the appropriate cells in the 2 X 2 table (Figure 3-3) are filled. Analysis is completed by comparing disease rates between the exposed and the unexposed in the study population.

It is also possible to modify the prospective approach, by using a **nested,** or design within a design approach. Nested designs are frequently used in large cancer or birth defect studies. In the prospective study, while the population is being characterized by exposure, a small group of especially selected people are set aside for further study at some point in the future (Frisch et al. 1990). For example, the National Cancer Institute, the National Institute of Environmental Health Sciences and the Environmental Protection Agency are collab-

		DISEASE		
		PRESENT	ABSENT	
EXPOSURE	PRESENT	a	b	a + b
	ABSENT	c	d	c + d

Incidence in the exposure present group = $\dfrac{a}{a+b}$

Incidence in the exposure absent group = $\dfrac{c}{c+d}$

FIGURE 3-3. Prospective Study and Related Calculations.

orating on a study of non-hodgkins lymphoma in pesticide applicators. The study design is prospective—over a five year period, the investigators will gather data (exposure and health) on a cohort of almost 200,000 people. This cohort will be followed for 10 or more years to see how many of the applicators contract non-hodgkins lymphoma. Nested within this cohort will be an especially selected group of applicators (approximately 15,000) who will be asked to provide blood to be stored until such time as the donor acquires cancer. At that time the donors blood, and control blood, will be analyzed for biomarkers of exposure and disease (as yet undetermined) that may turn out, upon analysis, to be predictive of the disease.

The person-year concept (100 persons exposed = 100 person years of exposure) is an advantage in prospective studies (by permitting a reduction in the number of study subjects, while increasing the number of years exposure). However, the long follow-up required before disease onset can still make this approach very costly (Mausner and Bahn 1974).

The Historical Prospective Design. The prospective and retrospective designs can be modified to include important characteristics from both designs. This modification is found in the nonconcurrent or historical prospective study. In this design, a study group (disease free) is identified on the basis of a **past exposure and followed forward in time until disease or death occurs.** How far back in time the investigator goes to establish exposure is a function of the availability of an exposure history, the latency period of the disease being investigated, adequate employment or other historical records, and the number of cases required to test the hypothesis. It is even possible to follow this population forward from the present, if some of the cohort remain alive. There are several criticisms associated with historical cohort studies: Subjects lost to follow-up; poor exposure classification;and inappropriateness of the comparison group data (for example, national data). However, historical cohort findings may be defended on the basis of consistency, either within the data (for example, dose-response across exposure levels; cancers that are biologically plausible; etcetera.) or through consistency of agreement with other studies.

Several software programs are available commercially (Hill 1972; Monson 1974; Marsh and Preminger 1980; Waxweiler et al. 1983) for use in historical prospective designs. The software takes data from historical cohorts after 1940 and performs the tedious work involved with arriving at the **expected** values needed for statistical analyses. Comparison data are available within the software for general mortality and cancer occurrence. A variety of supportive materials have been developed for use with these systems for classification of occupational exposures which may be adaptable to well characterized environmental settings (Hoar 1980).

The Case-control Design. The case-control design is sometimes identified as a retrospective study because of the use of historical exposure data. It is sometimes called case-comparison so that the design will not be confused with controls in experimental studies. The case-control design differs from the retrospective design previously discussed because it starts with a disease and **looks back in time to associate the disease with exposure**. The retrospective cohort, on the other hand, starts with a historical exposure and looks from the past to the present to associate exposure to disease. A benefit of the case-control design is that it can be employed to **evaluate a number of exposures in relation to one disease** (Monson 1980). The key to a successful case-control study is to establish the best disease history possible since misclassification occurs when disease cases are not accurately characterized. Misclassification influences are addressed as a part of the design of the research study (for example, use of specific case definitions or exposure criteria, blinded data collection, etcetera). The accurate assignment of subjects to diagnostic groups and to exposure categories is a substantive issue. While the proportions for these misclassifications can rarely be estimated, steps to minimize effects should be part of study planning and conduct.

Traditionally, information on exposure has been gathered through work histories; personal interviews; and occasional monitoring of the ambient environment. Often, surrogates (for example, liters of water drunk per day per unit of exposure or living distance from a toxic waste disposal site) have been used to estimate exposure. For diseases involving children, surrogate respondents (usually a parent) are requested to provide an exposure history. When working with a decedent case, survivors are queried as to their knowledge of the decedent's exposure history. With the advent of risk assessment as a discipline, however, and the requirements by risk managers for more specific exposure information, the use of surrogates is rapidly becoming less acceptable. Highly sophisticated monitoring of the ambient environment (although costly) is becoming *de rigueur* in the risk management process. The use of biological markers of **exposure and effect** (for a more complete discussion on biological markers see Chapter 8), are also expected to greatly enhance the utility of epidemiologic studies in the risk assessment process (Perera 1982). Although investigators may use **incident, prevalent,** or **mortality** cases, for case-control studies, it is generally believed that incident cases are better able to provide an accurate exposure history. The selection of controls is potentially the most important aspect of designing a case-control study. Controls must be concurrent with the cases, and be representative of the population from which the cases originate. Cases must be representative of the population of interest, and medical records should be utilized whenever possible to ensure that cases are accurately identified and classified. For this reason, case-control studies are often conducted in a hospital setting, although data from disease registries, and surveillance systems may also be used to provide cases. Regardless of the population based used, a case-finding system is needed for identifying all eligible cases.

Controls are usually selected from either the general (reference) population, or from hospital records. Controls from the general population should reflect the population in the study area from which the cases are derived. If the controls are selected from hospital records, they should reflect the factors within the community that caused the case to be admitted to that particular hospital (for example, area of residence, ethnic group, socioeconomic status). Some investigators use other types of disease cases as controls (for example, a type of cancer that is clearly not related to the cancer being investigated)(Calle 1984; Pearce and Checkoway 1988; Wacholder and Silverman 1990). However, when other disease cases are used, every effort should be made to use a variety of diagnoses in order to ensure that the controls properly reflect the characteristics of the reference population.

Occasionally, investigators will choose to match the exposed and controls on selected factors that may otherwise **confound** the findings (for example, age, sex, race, occupation, smoking history). The objective of epidemiologic analysis is to compare disease rates between exposed and unexposed persons for the purpose of determining the etiology of a disease. This can only be accomplished if the exposed and unexposed are comparable in all aspects, **except for the exposure.** In order to assure comparability, the cases and controls are **matched** for characteristics believed to influence the distribution of exposure. Age, sex, race, socioeconomic status, and occupation are all believed to have some influence on the exposure issue. Consequently, these factors are often used to match cases and controls. Obviously, once a factor is used for matching, its etiologic role can no longer be investigated because the cases and controls are not different in respect to that particular factor.

Matching can be for the whole study group or for individuals. Individual matching may be "1 to 1", or "k to 1,"
which means that more than one comparison person is selected for each study subject. **Confounding bias** occurs when a study variable is associated with the exposure and independently is a cause of the disease (Monson 1980). Once a factor is matched, however, its etiologic role cannot be investigated because the cases and controls are no longer different in respect to that particular factor. There are other methods to control for confounding. For example, stratification and refined statistical techniques (such as, multivariate analyses) are used in the analysis phase to control for confounders. However, if more sophisticated statistical techniques are to be used to an advantage the appropriate data must be collected—-for example, a **more complete** exposure history, with categories ranging from "never" to "very high", rather than simply yes or no responses (Kahn 1983; Mantel and Haenszel 1959). Statistical consultation is highly recommended when planning a study using more advanced analytical methods.

Diagnoses become more complex, and exposure needs more specific (for governmental regulatory purposes) as chronic disease epidemiology has become more prominent. The case-control model is being used more often in chronic disease studies because it can be completed in a shorter period of time. It is more efficient when studying diseases that are relatively rare in the community (it is cheaper and faster to reach an adequate sample size), and it is considerably less costly than the prospective cohort approach.

In conducting case-control studies, particular attention must be paid to the integrity of the data: Coding (narrative and numerical), and data entry into the computer should always be accompanied by appropriate quality control checks. Considerable attention must also be given to assuring the quality of surrogate responses—-for example, checks for internal consistency within answers, and verify reported information through other sources (Pickle et al. 1983). Procedures have also been developed to evaluate the reliability of a respondent's reporting of symptomatology in an emotionally charged study (Roht et al. 1985).

The Cross-sectional Approach

The **cross-sectional** approach reflects a "look-and-see attitude" to data gathering. Data shown in Figure 3-4, reflect a cross-sectional observation of pesticide workers who are exposed to organophosphate insecticides and the level of **DEP** found in their blood. These are compared with unexposed controls taken from the **NHANES** sample. The fundamental difference between the cohort and cross-sectional designs is that exposure and disease status are determined simultaneously in the cross-sectional approach, and

	Diethyphosphate (DEP) Residue Values			
	Yes	No	Total	Rate
Pesticide Use	228	104	332	68.7
Controls	427	6467	6894	6.6

FIGURE 3-4. Cross-Sectional Data.

therefore the time sequence between exposure and disease cannot be inferred.

Data analysis in a cross-sectional study is similar to that of a cohort or case-control study (for example, disease rates can be compared between exposed and unexposed groups; the proportion of persons with exposure can also be compared between the diseased and non-diseased. Importantly, data from two cross-sectional studies completed on the same population at different points in time can be analyzed similarly to prospective cohort studies (Monson 1980).

However, in a cross-sectional study **it is impossible to determine whether exposure precedes or follows the onset of disease.** Consequently, this approach is used less frequently in community epidemiologic studies than in occupational studies where employment records may permit an exposure history to be determined. Although this approach is most often used to investigate acute conditions (for example, skin rashes, respiratory effects) it increasingly is used to study chronic conditions (for example, decreased spermatogenesis in workers exposed to **DBCP**; neurotoxic effects among persons exposed to contaminated water supplies).

A common modification of the cross-sectional approach occurs when investigators use the total group as the unit of measurement, rather than the individual (for example, no **personal** information on individual members of the study population is gathered). Study populations may be found in several environments, including but not limited to, work places, schools, homes, neighborhoods, screening programs, and disease registries. Available health data (very often mortality and/or morbidity data from cancer registries, or reproductive disease registries) are summarized by population factors that are known to be predictive of health status such as age, sex, race, social class, marital status, parity, maternal age, and birth weight. Study groups are delineated by some measure of exposure (for example, proximity to a copper smelter; living near a toxic waste site) of interest to the investigators. They are then sorted by a group characteristic (for example, zip codes, street addresses, and census tracts) and statistical analyses completed to describe possible environmental associations. By modifying the cross-sectional approach to apply statistics to groups rather than individuals, it is possible to study large numbers of people at reduced costs. Generally, the cross-sectional approach that relies on group data is considered to be an ecologic or hypotheses- generating study. The ecologic approach may also be applied with trend statistics, as measures of change over time or along an exposure gradient (Fuortes et al. 1990).

The principal advantages of the cross-sectional approach are its relatively low cost and rapid completion time. Its principal limitation, as previously noted, is its inability to determine whether exposure precedes disease. This design also has the potential for selection bias in classifying the exposed and control study groups. With ecologic or descriptive data, consistency with other findings and trends within the data are not sufficient to permit strong inferences. Such studies do, however, provide justification for more elaborate research activities. Potential errors from making inappropriate inferences based on ecologic data has been called the ecologic fallacy (Morgenstern 1982).

ESTABLISHING RISKS IN ENVIRONMENTAL EPIDEMIOLOGY

Typically, epidemiologists will estimate health risks by determining the rate of occurrence of the disease in the study population (this is termed the absolute risk (AR)). To compare risk in the study population with risk in the comparison population, epidemiologists determine incidence rates or ratios. Ratios are sometimes used rather than rates because of the difficulty in ascertaining numerator (disease) and denominator (population at risk) data among the exposed and unexposed groups. The incidence rate (or ratio) among the exposed group is compared to the incidence rate (or ratio) of the disease among the unexposed (comparison) group (this is known as the relative risk [RR]) as shown in Figure 3-5.

If the RR is equal to one, the risk of disease in an exposed person is considered to be the same as in an unexposed person. However, if the RR is greater than one, then the risk of disease is considered to be greater for an exposed person than for an unexposed person. A negative association is determined when the RR is less than one, and the exposure is considered to potentially have some protective effect. As an example of RR used in prospective studies, genetic mutations in workers exposed to the fumigant phosphine are compared with genetic mutations in non-exposed controls (Figure 3-6). The RR (5.1) indicates clearly that the exposed group of grain elevator workers is more likely than the controls to experience genetic

$$RR = \frac{\text{Risk of the disease in exposed individuals}}{\text{Risk of the disease in unexposed individuals}}$$

FIGURE 3-5. Relative Risk (RR).

		Genetic Characteristic	
		Mutations	No Mutations
EXPOSURE	PRESENT	a 50	b 20
	ABSENT	c 10	d 60

Incidence rate in exposed individuals $= \frac{a}{a+b} = \frac{50}{70} = .71$

Incidence rate in unexposed individuals $= \frac{c}{c+d} = \frac{10}{70} = .14$

Relative Risk (RR) $= \frac{.71}{.14} = 5.1$

FIGURE 3-6. Calculation of Relative Risk (RR).

mutations. Even without an appropriate denominator, risk can be estimated in case-control or retrospective studies if it can be assumed that the control group is representative of the general population; the cases are representative of all cases; and the disease is relatively rare in the population (Mausner and Bahn 1974). When the epidemiologist determines that the study population fits these criteria, an odds ratio (OR) can be used to estimate risk (Figure 3-7). The OR is so named because it presents the odds in favor of having the exposure with the disease present and with the disease absent, respectively.

Another important measure of risk used by epidemiologists is the attributable risk (AR) as shown in Figure 3-8.

The attributable risk (AR) is a measurement of the amount (percent) of the absolute risk that can be attributed to the exposure under investigation. To calculate AR, the level of risk in the exposed group is subtracted from the level of risk in the unexposed group (a common risk shared by both the exposed and unexposed groups) and divided by the level of risk in the exposed group. For example, in Figure 3-9, the data show the AR associated with exposure to phosphine, and genetic mutations among occupationally exposed workers.

In Figure 3-10, the data show the AR for smoking and lung cancer mortality. The attributable risk is important to risk managers regulating

		Genetic Characteristic		
		Mutations	No Mutations	
EXPOSURE	PRESENT	a 50	b 20	a + b
	ABSENT	c 10	d 60	c + d
		a + c	b + d	

$$OR = \frac{ad}{bc} \text{ or } \frac{\frac{a}{c}}{\frac{b}{d}} = \frac{(50)(60)}{(20)(10)} = \frac{3000}{200} = 15$$

FIGURE 3-7. Calculation of OR Using Genetic Mutation Data. (Rare event criteria not met.)

$$AR = \frac{\text{Incidence among exposed} - \text{Incidence among unexposed}}{\text{Incidence among exposed}} \times 100$$

FIGURE 3-8. Attributable Risk (percent).

Exposure	Ratio of Genetic Mutations
Phosphine Exposed	.71
Unexposed	.14
Relative Risk	$\frac{.71}{.14} = 5.1$
Attributable risk (AR)	.71 − .14 = .57
	$\frac{.57}{.71} = .80 \times 100 = 80.3\%$

FIGURE 3-9. Attributable Risk (1) Calculation.

	Annual Death Rates per 100,000
Exposure Category	Lung Cancer
Heavy Smokers	166
Nonsmokers	7
Measure of Excess Risk	
Relative Risk (RR)	$\frac{166}{7} = 23.7$
Attributable Risk (AR)	$166 - 7 = 159$
	$\frac{159}{166} = .957 \times 100 = 95.7$

FIGURE 3-10. Attributable Risk (2) Calculation. Source: Adapted from Doll & Hill 1956.

environmental exposure because it provides them with a basis for quantifying how much risk is attributable to a specific exposure. Once this information is available, the risk manager can estimate how much the risk will be reduced per unit of reduction in exposure.

MEASUREMENT ERROR

Accurate data are the cornerstone of epidemiologic studies. Consequently, in epidemiologic research, investigators must be aware of (1) the sources of measurement error in selecting study populations, and (2) the reliability and validity of the survey instruments that are the backbone of our information gathering system. In epidemiologic studies we attempt to draw inferences to entire populations. To accomplish this, we must either study 100 percent of the target (**reference**) population, or select an accurate **sample** of that population. We will simply be unable to draw correct inferences if our sampling procedure does not permit us to make accurate estimates of the characteristics of the target population (that is, the sample must be representative of the entire target population). There are two basic sources of **measurement error** in any epidemiologic study: **random** and **systematic** error.

Random Error (or precision) is the measure of random variation around a specific value. For example, if we have eight different people draw blood from persons who are exposed to an organophosphorus insecticide (OP) that depresses a blood enzyme (cholinesterase) from a population (with a known mean cholinesterase level of 0.74 Δ pH units per hour), we could observe the following cholinesterase values: **0.95, 0.94, 0.93, 0.81, 0.74, 0.73, 0.67, 0.50**. This dispersion of values around 0.74 reflects the random error (or **precision of measurement**).

The possibility of **systematic error or bias** occurs when attempting to measure an association between variables. The opportunity for bias to occur is somewhat greater in the retrospective than in the prospective epidemiologic study design. When it does occur, in either design, the result will be an error in the estimation of the strength of the association (Relative Risk).

Selection or Non-response Bias

Selection or non-responsebias occurs when there is **systematic error** resulting in noncomparable admission of diseased (or nondiseased) persons into the exposed (or unexposed) group. In fact, there must be a difference in the selection criteria between the two groups for selection bias to occur. From the cholinesterase data shown above, lets assume that in the selection of the workers, we systematically sampled a group of workers with **reduced** exposure to the OP. The observed mean would then be expected to be higher than the true mean of the total population (0.74 Δ pH units per hour): 0.94, 0.93, 0.90, 0.89, 0.85, 0.83, 0.82, 0.79. The location of the sampled mean (0.87 Δ pH units per hour) reflects the accuracy of the estimate, or systematic error (bias) resulting from a draw of blood from the least exposed workers. Clearly, if the nonparticipants have more exposure, and possibly more reduction in the ChE values than the participants, then serious bias is introduced into the selection of study subjects.

One method epidemiologists use to reduce selection bias is by having a 100 percent census of the study population from which to select a sample, and to ensure that the sample is completely randomized. Another method **(the double blind study)** is to keep both the scientist and the participant ignorant of the group to which the subject has been assigned. If only the scientist knows the assignment, this is known as a **single blind study.**

Observation (Information) Bias

If, while conducting a case-control study of the risk of bladder cancer among persons drinking chlorinated drinking water, more complete information on the consumption of water is obtained from the controls than from the cases, the estimate of people drinking chlorinated water will be low, while the estimate in the controls will be correct. Such observation bias would be expected to force the analysis toward the null hypothesis (that is, the hypothesis of no difference between study groups) leading to an OR of 1.0, or no effect.

Confounding Bias

In any study attempting to measure an association between exposure and disease, confounding bias occurs when **a study variable is associated with the**

exposure and independently is a cause of the disease. For example, in a study of coffee drinking and pancreatic cancer, smoking may play a confounding role since coffee drinkers tend to smoke more than non-coffee drinking controls. Controlling for smoking, which is associated with pancreatic cancer is critical to the analysis since smoking is more common among the coffee drinkers than the non-coffee drinkers. Consequently, if smoking is not controlled, it will appear that coffee drinking may be placing people at greater risk to pancreatic cancer, when much of the risk may, in fact, actually be attributable to smoking. Smokers in both groups may mask the effect that coffee actually has on the relationship with pancreatic cancer by driving the results toward the null (this is negative confounding). Confounding bias can generally be accounted for in either the design of the study or during data analysis. However, we can never control for all the potential confounders. If there is good reason to believe that real confounders exist, they should be considered in the analyses. If they are not, money and resources can be better spent elsewhere.

Berksonian Bias

In a hospital based case-control study, it is possible that a spurious statistical association will be obtained between and exposure and a disease because of factors associated with the admission practices of the hospital (hospital records are not always complete; physicians do not hospitalize all their patients with a particular disease; some patients die before admission). Or the spurious association may occur because the study participants do not accurately reflect the catchment area for all hospital admissions.

Data Reliability and Validity

Reliability is the extent of agreement between repeated measurements (that is, the instrument, questionnaire, or medical evaluation device gives consistent results when used more than one time on the same individual or on a group of individuals). Two sources of error impact on reliability. **Observer error** occurs when one or more observers report different findings on separate occasions. **Method error** (variation) occurs when the test instrument or materials provide inconsistent findings using the same protocol. Validity is defined by **sensitivity** (the probability that a sick person will be classified as sick), and **specificity** (the probability that a healthy person will be classified as healthy). It is useful, when trying to detect or predict the event in a population, to think of the distribution of subjects according to the presence or absence of the exposure and the presence or absence of the event (Figure 3-11). Sensitivity is then estimated as the ratio of the number of subjects

	Diseased	No Disease	
Test Disease Present	a	b	a + b
Test Disease Absent	c	d	c + d
	a + c	b + d	a + b + c + d

Sensitivity $\dfrac{a}{a+c} = \dfrac{\text{Number tested positive}}{\text{total with condition}}$

Specificity $\dfrac{d}{b+d} = \dfrac{\text{Number tested negative}}{\text{total without condition}}$

FIGURE 3-11. Sensitivity and Specificity.

positive for both the exposure and the event to the number of subjects with the event. Specificity is estimated as the ratio of subjects negative for both the exposure and the event to the subjects negative for the event.

In epidemiologic studies we are always concerned that a participant who is diseased will be identified as being not-diseased. This is known as a **false negative**. Occasionally, a person who truly does not have a disease is declared to be diseased; this is known as a **false positive.**

MORBIDITY AND MORTALITY MEASUREMENT

Standardized Mortality Ratio (SMR)

The SMR is the ratio of the number of deaths in a specific population to the number of deaths that would have been expected when the population was compared with the same mortality experience in a **standard** population (comparable in many respects to the study population except for exposure history). When using the SMR, person years are often used as the denominator.

Standardized Incidence Ratio (SIR)

The Standardized Incidence Ratio of two standardized rates that have been standardized to the exposed distribution. The numerator of the SIR is the crude rate in the exposed population, and the denominator is a weighted average of the category-specific rates in the unexposed population, weighted by the distribution of the exposed population (Rothman 1986). The data in Table 3-4 provide an example of SIR computation.

TABLE 3-4. SIR of Cancer Prevalence Among Love Canal Residents.

Observed cases of cancer:	142
Expected cases of cancer:	126
SIR = 142 / 126 = 1.12 (0.94, 1.32; 95% Confidence Limits)	

Proportional Mortality Ratio (PMR)

The PMR is the ratio of deaths for a specific disease among all deaths in an exposed group to the ratio of deaths for that disease among all deaths in an unexposed group. Proportional mortality ratio's are used when it is impossible to obtain denominators for calculating rates. The PMR ratio is an acceptable approximation of RR when (1) mortality is comparable in the study populations, (2) deaths are classified the same way in both populations, (3) the ascertainment of deaths is complete in both populations, (4) competing causes of death are the same in both populations, and (5) data needed for proper adjustments are available for both populations.

Proportional Incidence Ratio (PIR)

The PIR is a simple risk measure that compares the observed (O) number of events with the expected number (E) of events. The PIR computation is shown in Figure 3-12.

	Residence adjoining the canal	Residence outside of the area
Total number of pregnancies	79	125
The observed number of SAs	15	11
Proportion of pregnancies ending is SAs	19%	9%
PIR = 19% / 9% = 2.11[a]		

[a] The probability of doubling the risk for spontaneous abortions being a random finding is less than 5% (p < 0.05). (Frumkin and Kantrowitz 1987). Adjusted for age and parity. Respective p values were calculated with the Mantel-Haenszel one-sided chi square test.

FIGURE 3-12. Frequency of Spontaneous Abortion (SA) in Residents Living Near, and Removed from the Love Canal.

STATISTICAL CONSIDERATIONS

Too often, statisticians are brought in to an epidemiologic study at the time of data analysis and are asked to find something in the data. When sophisticated designs are necessary, expert statistical consultation should be employed at the beginning of the study (for example, the design phase). There are, however, several excellent basic statistics books available for review. We recommend Duncan et al. (1983) and Rimm et al. (1980) for a technical discussion on statistical methodology. In this chapter we are briefly going to discuss a few of the basic statistical issues encountered in the analyses of epidemiologic data: significance tests, confidence intervals, sample size, and statistical power.

Significance Tests and Confidence Intervals

Underlying all statistical tests is the concept of the **null hypothesis.** For tests involving comparisons between two or more groups, the null hypothesis states that there is no difference among the groups in the parameters being compared (that is, the null is consistent with the idea that the difference is actually the result of random variation in the data). To decide if the null is to be accepted or rejected, a test statistic is computed and compared with a critical value from a statistical table. When the test statistic exceeds the critical value, the null hypothesis is rejected and the difference in parameters is declared to be statistically significant. If the decision to reject the null hypothesis is made on the basis of the test statistic, it carry's with it the probability (called the significance level of the test) of being wrong (that is, there is no difference in the parameters). If we test at the 0.05 significance level (this is the level that is frequently used in testing scientific questions), there is a 5 percent chance that we will reject the null hypothesis when it actually should be accepted (Griffith and Duncan 1983).

We will frame this discussion within statements that are frequently found in the reporting of epidemiologic data:

1. the difference in sample proportions is statistically significant; and,
2. the difference in sample proportions cannot be explained by chance alone.

Garland et al. (1990) reported on cases of melanoma among active duty U.S. Navy personnel by job description. A total of 174 confirmed cases of melanoma were diagnosed among white male Navy personnel between the years 1974-1984. Occupational risk of melanoma was determined by categories of exposure to sunlight (indoor, outdoor, indoor/outdoor). The investigators calculated a SIR. "Two occupations showed statistically significantly high SIRs: Aircrew survival equipmentman, SIR = 6.8 ($p < 0.05$); and engineman, SIR = 2.8 (p < 0.05)." This is equivalent to saying that the difference in occurrence of melanoma in these occupational groups cannot be explained by chance alone (p < 0.05). The finding can also be explained by saying that the difference is significant at the 5 percent level.

In elementary statistics, we learn that different samples randomly selected from the same population will vary. The question with Garland et al., is whether the observed difference would be likely to occur in repeated samplings taken from populations with identical true percentages. The authors are saying that they would not expect a difference this large or larger to occur more often than five times in one hundred such samples (significannt at the 5 per cent level). However, there is always the possibility that the two populations were not really different, and that a large sampling error has occurred. If a sampling error has occurred, this difference might then be considered a Type I error or alpha error (α). **An alpha error occurs when a hypothesis is rejected that should have been accepted.**

However, if we believe that the difference in the aircrew survival equipmentman and engineman occupations is real, then it would follow that we would want to know how large the difference is. The information provided by any sample will depend on the sample size and the sample variation, and can be expressed as a **confidence interval or limit** for the difference in the population values.

As shown in Ahlbom and Norell (1984), the following methods are available to calculate confidence intervals (CI). With the 95 percent confidence limit, Garland et al. are saying that the true SIR in these two occupational groups lies between 2.5 and 14.7 for the aircrew survival equipmentman; and between 1.3 and 5.1 for the engineman. If a smaller Confidence Limit is desired, then larger sample sizes are called for. Small sample sizes lead to small expected values for SIR/SMR calculations. It is especially prudent to place confidence intervals on all risk estimates, and not provide p-values alone.

The calculation of the CI for the prevalence P of an event in a population can be calculated from the formula shown below, when D equals the number with the disease, and the N equals the total population:

$$P = \frac{D}{N}, \text{ a 95\% CI is calulated by}$$

$$P \pm 1.96 \left(\frac{P[1-P]}{N} \right)^{1/2}$$

The calculation of a 95 percent confidence interval to measure differences in the occurrence of disease can be calculated from the following formula, when D is the difference in the occurrence of disease between the exposed and unexposed populations, and 1.96 provides the level of confidence and I is the test value from the hypothesis test:

$$D(1 \pm 1.96/x)$$

The problem in the selection of a probability level (alpha level) is difficult in most experimental and observational epidemiologic studies since the requirements necessary for the most appropriate statistical theory are seldom met. In this regard, it would be worthwhile to consider the two major theoretical issues involved in statistical theory, which form the basis of almost all statistical practice and account for the application of statistical procedures based on the so-called **normal** theory. As described by Griffith and Duncan (1983) these are:

1. The sampling distribution of mean values is based on random samples from a **reference** population. The distribution of all possible mean values of repeated random samples of size n generates a derived population of mean values. The standard deviation of the derived population of mean values has a special name, the **standard error.** If the reference population has a mean value and a standard deviation %, then the derived population of mean values has a mean value and standard error of %/n.

   The mean of any particular sample is the sample estimate of the population mean. The sample standard deviation % is the sample estimate of the population standard deviation, and the sample standard error $s/\sqrt{n}$ is the sample estimate of the standard error of the derived distribution. This is **true** for every reference population regardless of its distribution (for example, normal, Poisson, binomial, chi- square, log-normal).

2. The principle embodied in the (weak) law of large numbers as expressed in the **central limit theorem**. This important theorem states (Duncan et al. 1983) that as a sample size increases (30 n), the derived distribution of mean values based on random samples of size n becomes **approximately normal**, with mean and standard error %n. This can be shown to happen quickly (for small n) even for very skewed distributions. This fact forms the basis for the wide spread, frequent application of normal theory tests (for example, the students's-t test, F tests, chi-square tests, multiple regression analyses, and multivariate analyses), which are known to be **robust** with respect to the violation of *inter alia* (the assumption of normality of the reference populations(s)). Fortunately, the tendency toward normality of estimates of mean values, governed by the **central limit theorem**, makes the use of normal theory tests especially the ubiquitous student's-t test, appropriate for comparing samples from different populations.

The rapid approach to normality of sample means from a general population does not depend on the size of the standard deviation with respect to the mean value, nor does the size of the sample standard deviation with respect to the sample mean indicate whether the application of, say, a t-test is appropriate. The critical relationship is that of the standard deviation to the sample size (that is, the **standard error**). Thus, a criterion that has intuitive appeal is that, for any given sample, the mean value must be at least twice the standard error if approximate normality is to be assumed.

Occasionally, with a very skewed distribution or, with data that are not otherwise **normally** distributed about the mean, one must utilize more complex statistical procedures. For example, to conduct statistical tests concerning the mean values of such distributions, the variable (for example, the number of occurrences) may be replaced by its square root; this is square root transformation. Another similar transformation that is sometimes used with environmental data is the exponential transformation or logarithmic scale. One approach is to use a **goodness-of-fit** test like those available in soft-ware statistical packages such as SAS (Statistical Analysis Software)(Hanrahan et al. 1990). The SMR can also be used for comparisons of this kind, so that age, race, sex and population change over time can be taken into account before the comparisons are made.

Statistical Power and Sample Size

The determination of the power of a test, and sample size requirements, are beyond the scope of this text. However, we will discuss the basic concept and use of sample size tables. To establish an appropriate sample size, an investigator should:

1. Consider the prevalence of the disease
2. Consider the level of risk to be considered meaningful (usually two-fold risks are a minimal criterion), and
3. Determine the level of Type-I and Type-II errors **(Type II error (β) occurs when a hypothesis is accepted when it should have been rejected)**

Commonly, the Type-I error is set at .01 and the Type II error is set at .05. Statistical power, while important, is somewhat of a subjective consideration. For example, an investigator may report no risk of disease associated with an environmental exposure, measured by a RR at the 1.5 to 2.0 level. Unfortunately, due to the size of the study group (too small), it was impossible to detect the **real** level of increased risk (5.0). Rather than considering such studies to have negative findings, they should be considered to be **inconclusive**. The term **negative study** is used to designate when an exposure/disease association does not exist, versus the term **inconclusive** which refers to a level of risk that cannot be addressed due to sample size limitations (Beaumont and Breslow 1981).

In attempting to determine an appropriate sample size, the investigator might feel that a difference in proportions as large as 20 percent would be necessary. In this regard, the investigator attempts to limit the chances of making a **false** claim of significance to not more than 1 percent (0.01), and establishes that it is important to have a 95 percent chance of detecting a difference as large as $P_1 = 20$ percent and $P_2 = 40$ percent. To determine the appropriate sample size at the 0.01 percent level, use Table 3-5, and follow directions:

> Enter the first column headed **"the smaller of the two percentages"**, on the row for 20 percent and read in the fourth column $(P_1 - P_1) = 20$ percent to find the size of **each** sample, n_1 and n_2, should be 184. If the investigator is willing to accept a greater probability (0.05) of making a Type I error using this approach refer to Table 3-6. With P_1 and P_2 remaining at 20 percent and 40 percent, respectively, the sample size requirements are reduced to n_1 and $n_2 = 134$. It is also clear, that if the size of the difference is increased, the required sample size is reduced.

The P Value of Significance ($P < 0.05$, $P < 0.01$)

Often, epidemiologic studies will report findings as a P value. This value expresses the probability that a difference as large as that observed in the study would occur by chance alone (referred to as the significance level of the test above). If we see the statement $P < 0.05$, then there is less than 5 chances in 100 that the difference observed would have occurred by chance

TABLE 3-5. Approximate Sample Sizes Necessary in Each of Two Groups Corresponding to Various Population Percentages in Order to Ensure 95 percent Power of the Two-Tail Test at the 1 Percent Level.

Smaller of Two Population Percentages	\multicolumn{9}{c}{Differences in Two Population Percentages}
	5	10	15	20	25	30	35	40	50	60
5	985	318	170	111	79	60	48	39	27	19
10	1,555	451	226	139	96	71	55	44	29	21
15	2,054	567	273	164	111	80	61	48	31	21
20	2,481	665	313	184	122	87	65	50	32	22
25	—	—	—	—	—	—	—	—	—	—
30	—	—	—	—	—	—	—	—		
35	—	—	—	—	—	—	—			
40	—	—	—	—	—	—				
45	3,550	887	392	217						
50	3,550	879								

Adapted from Fox et al. 1970.

TABLE 3-6. Approximate Sample Sizes Necessary in Each of Two Groups Corresponding to Various Population Percentages in Order to Ensure 95 percent Power of the Two-Tail Test at the 5 Percent Level.

Smaller of Two Population Percentages	\multicolumn{10}{c}{Differences in Two Population Percentages}
	5	10	15	20	25	30	35	40	50	60
5	719	231	124	80	58	44	34	28	19	14
10	1,135	329	164	101	70	52	40	32	21	15
15	1,498	413	199	119	80	58	44	34	22	15
20	1,810	485	228	134	89	73	47	36	23	15
25	—	—	—	—	—	—	—	—	—	—
30	—	—	—	—	—	—	—	—	—	—
35	—	—	—	—	—	—	—	—	—	—
40	2,538	641	286	160	101	70	50	38	23	
45	—	—	—	—						
50	2,590	641								

Adapted from Fox et al. 1970.

alone (that is, the probability that **random variation** accounts for the difference is very small, and that the difference is considered to be statistically significant).

Determining whether the *P* value is acceptable, versus being simply an artifact of the statistical methods employed in the analysis, may prove to be troublesome. **Caution is urged in the use of statistical significance. Statistical significance does *not* mean causally associated. It does mean that the recognized association has stability. In non-experimental studies the interpretation of the *P* value must be made with great care.**

Finally, the appropriateness of particular statistical models for testing the data (adjusting changing rates over time) can become very complex. In epidemiologic studies, certain alpha error levels (usually .05 or .01) are chosen, and all statistical tests are referenced to the chosen value. A problem occurs when multiple comparisons are made on a large data set from the same group of study subjects (Garland et al. compared cancer risks in 20 occupational groups and 176 total cases). In such a case, the alpha value may in fact be increased over the chosen value, and spurious conclusions may arise.

To guard against this happening during multiple contrasts, the customary statistical strategy is to make stringent demands on the p value required for **significance**. Instead of being set at the customary value of .05, the alpha level is substantially lowered. Although statisticians do not agree on the most desirable formula for determining this lower boundary, a frequently used procedure is to divide the customary alpha level by **k**, where **k** is the number of comparisons (this is known as the ***Bonferroni procedure***). The Bonferroni adjustment divides the desired, over-all p-value (usually $p < 0.01$ or 0.05) by the number of statistical tests to be performed; this step determines what level of significance one test must achieve for the entire analysis to maintain a given significance level (i.e., $p < 0.05$). Thus, in a study containing at least 20 comparisons, the decisive level of alpha would be set at no higher than .05/20 or .0025.

DETERMINING CAUSAL INFERENCES

In epidemiology, the determination of a true causal relationship is rare because eidemiologists are usually working with a complex set of exposures and outcomes that make it especially difficult to establish causation. For example, there are several environmental agents that are believed to cause leukemia, including radiation, and benzene.

If we hold to the position that there must be a one on one relationship between the factor under study and the disease (that is, whenever the factor is present so is the disease), we will likely never establish a causal relationship. So the question arises: "How do epidemiologists establish a causal relationship given the complex set of circumstances that they work under?" You will recall that in Chapter 2 we discussed a series of postulates developed by Robert Koch to determine a causal relationship in the infectious diseases. Clearly, Koch's postulates are not readily applied to the epidemiology of chronic diseases. We couldn't for example, reproduce the disease from culture into a healthy subject to see if the disease occurs. Also, attempting to extrapolate across species from laboratory animals to man is fraught with problems (for example, different dose-response relationships; metabolic properties are very different between species).

Yet, epidemiologists are regularly called upon to attribute causation to some disease/exposure relationship. In 1964, an advisory committee for the Surgeon General of the United States Public Health Service, cognizant of the difficulty in establishing causal relationships in chronic disease, developed criteria that, when properly applied, may be used to establish a causal relationship. A variation of those criteria follow:

1. **Strength of the association.** If the disease under study is found in all the exposed persons, but in none of the unexposed, a one to one relationship exists between the disease and the exposure, and causality may be inferred. However, in the real world most diseases have multiple etiology, so one to one relationships are unlikely to occur.
2. **Specificity of the association.** If exposure to the environmental pollutant under study is also associated with causation in other diseases, specificity

is likely to be weak. Also, the greater the number of causal agents producing a given disease, the weaker the specificity of the association.
3. **Dose-response relationship.** If exposure to an environmental pollutant can be said to be of causal importance to a disease, then the risk of developing the disease should be related to the degree of exposure to that pollutant. A dose-response relationship should occur.
4. **Consistency of association.** To establish consistency, epidemiologists should determine the distribution of the environmental pollutant (the etiologic factor) by as many characteristics (for example, gravida, age, sex, race, occupation) as possible. The distribution of the disease by the same population characteristics should be determined, and the distribution patterns analyzed to see if they properly overlap.
5. **Temporal relationship.** If a toxicant is to be considered the cause of the disease, exposure must have occurred prior to the disease. An example of this problem can be found in developing a causal relationship between smoking and lung cancer. Clearly, the lack of specificity is troubling (smoking is also associated with emphysema, heart disease, and bladder cancer). However, since tobacco is a mixture of chemicals, and many of the components are carcinogenic, including benzene, this tends to diminish the concern over weak specificity. Another factor that tends to reduce concern over weak specificity is biological commonality among cells and molecular structure. All cells in an organism have some similarity among the cell structure, either in the intracellular makeup, or in the chemistry of the cell. Thus it is not unreasonable to expect that a single agent could be responsible for disease occurring in different organ sites.
6. **Biologic Plausibility.** Occasionally, a factor and a disease are associated only because they are both related to some underlying condition. An example of this is the association between smoking and yellow index finger, and smoking and lung cancer. Clearly, there is no reason to believe that a yellow index finger causes lung cancer. However, there is good reason to believe that smoking causes lung cancer and yellow index finger (that is, smoking is a confounder). Heavy smokers often have a yellow index finger; they also have excess cancer of the lungs. Clearly, the yellow index finger is directly associated with smoking, but only indirectly (not a real association) associated with lung cancer. Thus, the association between yellow index finger, smoking and lung cancer may in fact be an **indirect, but non causal, association.** Epidemiologists, while searching for causality in their studies should always be concerned that there is **biological plausibility** in determining a causal relationship. The biological nature of the disease must play a role in any causal decision on the event. One might ask, given the pathological mechanism of the disease, "Is it biologically plausible that lung cancer could be caused by a yellow index finger?"

As a contrast, consider the question "Is it biologically plausible that lung cancer could be caused by inhaling tobacco smoke?"

SUMMARY

In this chapter we have provided an overview of the primary methods used in epidemiologic studies, and discussed many of the technical issues with which an epidemiologist must be familiar to successfully conduct an environmental epidemiology study. To briefly encapsulate the epidemiologic process: Initially a hypothesis is made about an observation in the population. We remember from Chapter 2 that Snow made an observation about the source of drinking water and cholera. Descriptive epidemiologic studies may also be used to develop observations on disease in selected groups, using identified environmental exposures. These observations are then formulated as hypotheses. The hypotheses are tested by various epidemiolgic designs. Risk factors are identified, and intervention activities are begun to reduce the occurrence of disease.

ASSIGNMENTS

1. Define Descriptive and analytic epidemiology.
2. Describe the important differences between a cross-sectional study and a cohort study.
3. Define bias and describe one way it may be controlled in the selection of study subjects.
4. Define Relative Risk and describe how it would be used in establishing risk to an exposed population.
5. Define systematic error. What design flaws will result in systematic error in sample selection?
6. Define biological plausibility in the process of establishing causal inference in a study.
7. Define validity and its place in epidemiologic studies.
8. Describe a study situation in which confounding bias might occur, and how it might be controlled.
9. Consider how the issues of sample size, potential exposure, and latency period might complicate the interpretation of cancer incidence data.
10. How would you calculate the incidence of pesticide poisonings and a 95 percent confidence interval in a population (n = 1229) of agriculture workers that has experienced 12 pesticide poisonings within the past year.

GLOSSARY

Comparison groups: In some disciplines (for example, biology or chemistry), true "control" groups (those without any exposure other than that of the study) may be possible. People may not be so manipulated; thus a true control group is never actually attained in epidemiology. Rather, persons that represent the general population or background situation are used as a comparison to the study group.

Confounder: This term refers to a factor that is known to be associated with the occurrence of a health event. As an example, most health events vary by age, race, sex, etcetera. If such a known risk factor is not taken into account, the study results may be "confounded" (mixed up), so that an effect due to the study exposure may not be distinguishable from the already known factor (for example, race, sex). In the study analysis, the influence of a confounder can be separated from the effects of the study factor; also, two factors can be analyzed together for combined effects.

Exposure: This term may refer to an ambient environmental factor (for example, air pollution), to a factor in the individual's environment (for example, smoking and diet) or to personal characteristics (for example, age, race, and sex). In general, an exposure is considered as a precedent factor associated with a likelihood of one experiencing some health event or endpoint.

Endpoint: Usually this term refers to an undesirable health event, as the occurrence of disease or death. Yet it could represent a beneficial event as well (for example, recovery from an illness or healing of a wound). In general, an endpoint is the health consequence of some exposure. To view the exposure/endpoint relationship as being a "cause and effect" sequence is unrealistic in epidemiology. Exposure-endpoint associations imply a risk relationship, not certain causation.

Epidemiologic reasoning: A three-step sequence is used in epidemiologic reasoning: (1) determination of an association between an exposure and an endpoint, (2) formulation of a biologic inference (that is, a hypothesis) about that relationship, and (3) testing the hypothesis.

Nested Design: Nesting combines two study designs to get the best of both choices: more data, quicker, and at lower costs.

Person-years: For study designs where subjects are followed through time, each person contributes a person-year for each year of participation. For example, a group of 100 persons, studied from January 1, 1970, to December 31, 1979 (ten years), contribute 1000 person-years to the study.

Risk: This term expresses the relationship between the health experience observed for the study group and what would have been expected on the basis of the comparison group's health experience. When this value is expressed as a ratio (risk = observed/expected), it is a measure of the strength of association that exists between an exposure and an endpoint. It may be shown as a decimal fraction (1.5) or multiplied by 100 to give a percentage (150%). Risk may imply either an adverse (>1.0) or a beneficial (<1.0) relationship; no risk is shown by a value of 1.0.

Statistical techniques: The general premise behind statistics is the use of information gained from a sample that is a subset of a larger group in order to make statements about that larger group. This larger audience of persons (whom the sample represents) is the population of interest. Statistical methods are important to epidemiology; they are used to help distinguish between a real association and one that results from chance. These methods are influenced by the number of subjects used for the calculations.

RECOMMENDED READING

Duncan, R.C., Knapp, R.G., Miller, M.C. III. 1983. *Introductory Biostatistics For The Health Sciences.* New York: John Wiley & Sons.

Kleinbaum, D.G., Kupper, L.L., and Morgenstern, H. 1982. *Epidemiologic Research: Principles and Quantitative Methods* Belmont, California: Wadsworth Publishing.

Rothman, K.J. 1986. Modern Epidemiology. Boston/Toronto: Little, Brown and Company.

REFERENCES

Ahlbom, A., and Norell, S. 1984. *Introduction to Modern Epidemiology.* Chestnut Hill, MA: Epidemiology Resources, Inc.

Baker, D. B., Greenland, S., Mendlein, J., and Harmon, P. 1988. "A Health Study of Two Communities Near the Stringfellow Waste Disposal Site." *Archives of Environmental Health* 43(5):325–34.

Beaumont, J. J., and Breslow, N.E. 1981. "Power Considerations in Epidemiologic Studies of Vinyl Chloride Workers." *American Journal of Epidemiology* 114(6):725–34.

Calle, E. E. 1984. "Criteria for Selection of Decedent Versus Living Controls in a Mortality Case-Control Study." *American Journal of Epidemiology* 120:635–42.

Dawber T. R., Kannel W. B., Lyell L. P. 1963. "An Approach to Longitudinal Studies in a Community: The Framingham Study." *Ann. N.Y. Acad. Sci.* 107:539–56.

Doll, R., and Hill, A. B. 1956. "Lung Cancer and other Causes of Death in Relation to Smoking. A Second Report on the Mortality of British Doctors." *Br. Med. J.* 2:1071.

Duncan, R. C., Knapp, R. G., Miller, M. C. III. 1983. *Introductory Biostatistics for the Health Sciences.* New York: John Wiley & Sons.

Fox, J. P., Hall, C. E., and Elveback, L. R. 1970. *Epidemiology: Man and Disease.* London: MacMillan Co.

Fuortes, L., McNutt, L. A., and Lynch, C. 1990. "Leukemia Incidence and Radioactivity in Drinking Water in 59 Iowa Towns." *American Journal of Public Health* 80(10):1261–2.

Frisch, J. D., Shaw, G. M., and Harris, J. A. 1990. "Epidemiologic Research Using Existing Databases of Environmental Measures." *Archives of Environmental Health* 45(5):303–7.

Garland, F. C., White, M. R., Garland, C. F., Shaw, E., and Gorham, E. D. 1990.

"Occupational Sunlight Exposure and Melanoma in the U.S. Navy." *Arch Environ Health* 45:261–267.

Gordon T., Kannel W. B. 1970. *The Framingham, Massachusetts Study: Twenty Years Later. In the Community as an Epidemiologic Laboratory: A Casebook of Community Studies,* ed. Kessler, I. I., Levin, M. L., pp. 123–46. Baltimore: The Johns Hopkins Press.

Griffith. J., Duncan, R. 1983. "An Assessment of Fieldworker Occupational Exposure to Pesticides in the Florida Citrus Industry." In: *National Monitoring Study: Citrus, Vol. III.* Miami, Florida: University of Miami Press.

Hanrahan, L. A., Mirkin, I., Olson, J., Anderson, H., and Fiore, B. 1990. "SMRFIT: A Statistical Analysis System (SAS) Program for Standardized Mortality Ratio Analysis and Poisson Regression Model Fits in Community Disease Cluster Investigation." *American Journal of Epidemiology* 132-Supplement: S116–S122.

Hennekens, C. H., and Buring, J. E. 1987. *Epidemiology in Medicine* Boston: Little, Brown and Co.

Hill I. D. 1972. "Computing Man-Years at Risk." *British Journal of Preventive and Social Medicine* 26:132–43.

Hoar, S. K., et al. 1980. "An Occupational and Exposure Linkage System for the Study of Occupational Carcinogenesis." *Journal of Occupational Medicine* 22:722–26.

Howe, H. L. 1988. "A Comparison of Actual and Perceived Residential Proximity to Toxic Waste Sites." *Archives of Environmental Health* 43:415–19.

Kahn, H. A. 1983. *An Introduction to Epidemiologic Methods.* New York: Oxford University Press.

Kessler, I. I., Levin, M. L. 1970. *The Community as an Epidemiologic Laboratory: A Casebook of Community Studies.* Baltimore and London: The Johns Hopkins Press.

Kleinbaum, D. G., Kupper, L. L., and Morgenstern, H. 1982. *Epidemiologic Research: Principles and Quantitative Methods.* Belmont, California: Wadsworth Publishing.

Lilienfeld, A. M., and Lilienfeld, D. E. 1980. *Foundations of Epidemiology 2nd Edition.* New York: Oxford University Press.

Mantel, N., and Haenszel, W. 1959. "Statistical Aspects of the Analysis of Data from Retrospective Studies of Disease." *Journal of the National Cancer Institute* 22:719–48.

Marsh, G. M. and Preminger, M. 1980. "OCMAP: A User Oriented Occupational Cohort Mortality Analysis." *American Statistician.* 34:245–46.

Mausner, J. S., and Bahn, A. K. 1974. *Epidemiology: An Introductory Text.* Philadelphia, London, Toronto: W. B. Saunders Company.

Monson, R. 1980. *Occupational Epidemiology.* Boca Raton, Florida: CRC Press Inc.

Monson, R. R. 1974. "Analysis of Relative Survival and Proportional Mortality." *Computing and Biomedical Research* 7:325–32.

Morgenstern, H. 1982. "Uses of Ecologic Analysis in Epidemiologic Research." *American Journal of Public Health* 72:1336–44.

National Center for Health Statistics (NCHS), Public Health Service, Hyattsville, MD. U.S. Department of Health and Human Services. 1980. *Plan and Operation of the Health and Nutrition Examination Survey. A Description of a National*

Health and Nutrition Examination Survey of a Probability Sample of the U.S. Population 1-74 Years of Age.
Pearce, N., and Checkoway, H. 1988. "Case-Control Studies Using Other Disease As Controls: Problem of Excluding Exposure Related Diseases." *American Journal of Epidemiology* 127:851–6.
Perera, F., and Weinstein. 1982. "Molecular Epidemiology and Carcinogen-DNA Adduct Detection: New Approaches to Studies of Human Cancer Causation." *Journal of Chronic Disease* 35:581–600.
Pickle, L. W., Brown, L. M., and Blot, W. J. 1983. "Information Available From Surrogate Respondents in Case-Control Interview Studies." *American Journal of Epidemiology* 118(1):99–108.
Riggan, W. B., Van Bruggen, J., Acquavella, J. F., Beaubier, J., Mason, T. 1983. *U.S. Cancer Mortality Rates and Trends 1950–1979, NCI/EPA. Interagency Agreement on Environmental Carcinogenesis.* Washington, D.C. U.S. Government Printing Office EPA publication no. 600/1-83-015 a,b,c.
Rimm, A. A, Hartz, A. J., Kalbfleisch, J. H., Anderson, A. J., Hoffmann, R. G. 1980. *Basic Biostatistics in Medicine and Epidemiology.* New York: Appleton-Century-Crofts.
Roht, L. H., Vernon, S. W., Weii, F. W., et al. 1985. "Community Exposure to Hazardous Waste Disposal Sites: Assessing the Reporting Bias. *American Journal of Epidemiology* 122:418–33.
Rothman, K. J. 1986. *Modern Epidemiology.* Boston/Toronto: Little, Brown and Company.
Schechter, M. T., Spitzer, W. O., Hutcheon, M. E., et al. 1989. "Cancer downwind From Sour Gas Refineries: The Perception and the Reality of an Epidemic." *Environmental Health Perspectives* 79:283–90.
U.S. Department of Health, Education and Welfare. *Smoking and Health: Report of the Advisory Committee to the Surgeon General of the Public Health Service.* Public Health Service publication no. 1103. Washington, D.C. Government Printing Office, 1964.
Wacholder, S., and Silverman, D. T. 1990. "RE: Case-control Studies Using Other Disease as Controls: Problems of Excluding Exposure Related Cases." (Letter). *American Journal of Epidemiology* 132:1017–18. [Note also Authors Reply]
Waxweiler, R. J., Beaumont, J. J., and Henry, J.A. 1983. "A Modified Life-Table Analysis for Cohort Studies." *Journal of Occupational Medicine* 25:115–24.

4
Disease Clusters

Tim E. Aldrich, Wanzer Drane, and Jack Griffith

OBJECTIVES

This chapter will:

1. Review the major issues involved in the investigation of clusters (small groups of disease).
2. Discuss the basic concept of cluster identification and analysis.

BASIC STUDY DESIGNS

A basic premise of epidemiology is that disease distributes predictably in human populations (Doll 1981). From this perspective, patterns of disease are studied to identify groups who are experiencing greater rates of disease occurrence than others. This reasoning is along the traditional lines of person, place, and time as expressed in Chapter 3. Characteristics of person (age, race, sex) may identify higher disease occurrence groups. Place and time may signal changes in disease occurrence that concur within a geographic region (for example, county, zip code) or during a particular period of time (for example, 1980–85).

When we observe that disease occurrence is not uniform (for example, there are differences in the pattern of disease occurrence between population groups, geographic regions, or over time periods), then an epidemiologist attempts to identify a pattern of factors or events (determinants) that may explain the observed distribution. This is the sequence of epidemiologic reasoning:

1. Observe a statistical association (nonrandom pattern).
2. Develop a hypothesis about why the disease is distributed in that manner.

3. Test the hypothesis related to the determinant of the observed distribution.

The focal step, observation of disease patterns, is the reason that much of epidemiology is considered to be an observational science. When epidemiologists test hypotheses they do so through carefully selected observations and by comparing disease experience between groups. Examples of such exploratory analyses using epidemiologic data are **subgroup analyses** and **cluster studies**.

Subgroup Analyses

Subgroup analyses are straightforward comparisons within a larger data set. For example, comparing brain cancer rates among children to those of other age groups. However, the procedures for detecting excess disease occurrence within a subset of a population quickly become intricate when small numbers of incidents are to be studied, or if there is a wish to study disease patterns over time. Historically, such complex statistical analyses have been a deterrent to the analysis of disease cluster data, and conventional biologic reasoning and practical judgment were espoused instead (Caldwell and Heath, 1976; Caldwell 1989).

Cluster Studies

Cluster studies are quite another matter, and one that may be very controversial (Editorial 1990; Garfinkel 1987). Most states have programs for responding to reports of cancer clusters (Warner and Aldrich 1988). The heart of the process for evaluating disease cluster reports is determining whether case occurrence within a certain location is increased over the expected numbers. We usually look at the spatial pattern of cases within a defined geographic area, the time pattern for the occurrence of cases, and in some cases the space-time pattern for any evidence that the distribution of cases is not random (and could potentially be the result of a hazardous environmental exposure). However, most disease cluster reports are not associated with increased occurrence, and even those that involve increased disease rates are usually not relatable to an environmental point source.

Because of these factors, evaluating disease cluster reports is sometimes disparaged (Rothman 1990; Neutra 1990). However, over the years, health officials that responded to public outcry concerning reports of highly publicized disease clustering have been largely responsible for developing some central disease registries, and the evolution of population-based data that central registries can provide (Houk and Thacker 1987; Aldrich et al. 1990).

In any case, cluster studies are small area investigations and very small numbers of cases are usually involved (Thacker 1989).

These small numbers make the data analyses very subtle. A variety of specialized statistical methods have been developed for disease cluster analyses and computer software (***CLUSTER***) is available to assist with these analyses (Guidelines 1990; Aldrich and Drane 1990). It is important to understand that the application of statistical techniques is not a simple solution for assessing disease clusters. Statistical techniques are simply one more tool in the armory of epidemiologists. Biologic reasoning must be applied to interpreting a cluster report, as well as the use of statistical analyses. Statistics may, however, help to prioritize a number of cluster reports for investigation, or assist in making a decision not to investigate what is clearly not a cluster.

KEY QUESTIONS WHEN CONSIDERING A CLUSTER REPORT

A protocol for responding to disease cluster reports has been developed that is based on protocol components from several different states (Figure 4-1). Using this approach, slightly over half of the cluster reports may be closed at the level of reviewing data for proximal populations (for example, the surrounding county). Another large proportion of reports may be eliminated after a close review of the cases' clinical data (Heath 1990).

The use of a statistical battery (***CLUSTER*** software) for screening disease cluster reports is applied at the level of the protocol (Figure 4-1) entitled "Primary Evaluation of Proposed Cluster." Only 5 to 10 percent of all the cluster reports that an agency receives will ever reach the protocol step "Possible Clustering" (Table 4-1). For completeness, it is imperative to note that all communities in which disease clusters are reported should be offered on-site meetings and educational information in order to be responsible to citizen concerns (Fiore et al. 1990). Responding to disease cluster reports

TABLE 4-1. Distribution of Unusual Outcomes of Disease Cluster Investigations (Referring to Steps in Figure 4-1).

Specific Loop in Flow Chart	Closure for a Proportion of Investigations
Proximal and literature review	50–60%
On-Site interviews (chart reviews or interviews) (may include non–case-data as well)	20–30%
Primary evaluation of the proposed cluster	10–20%
Possible clustering (extensive studies)	5–10%

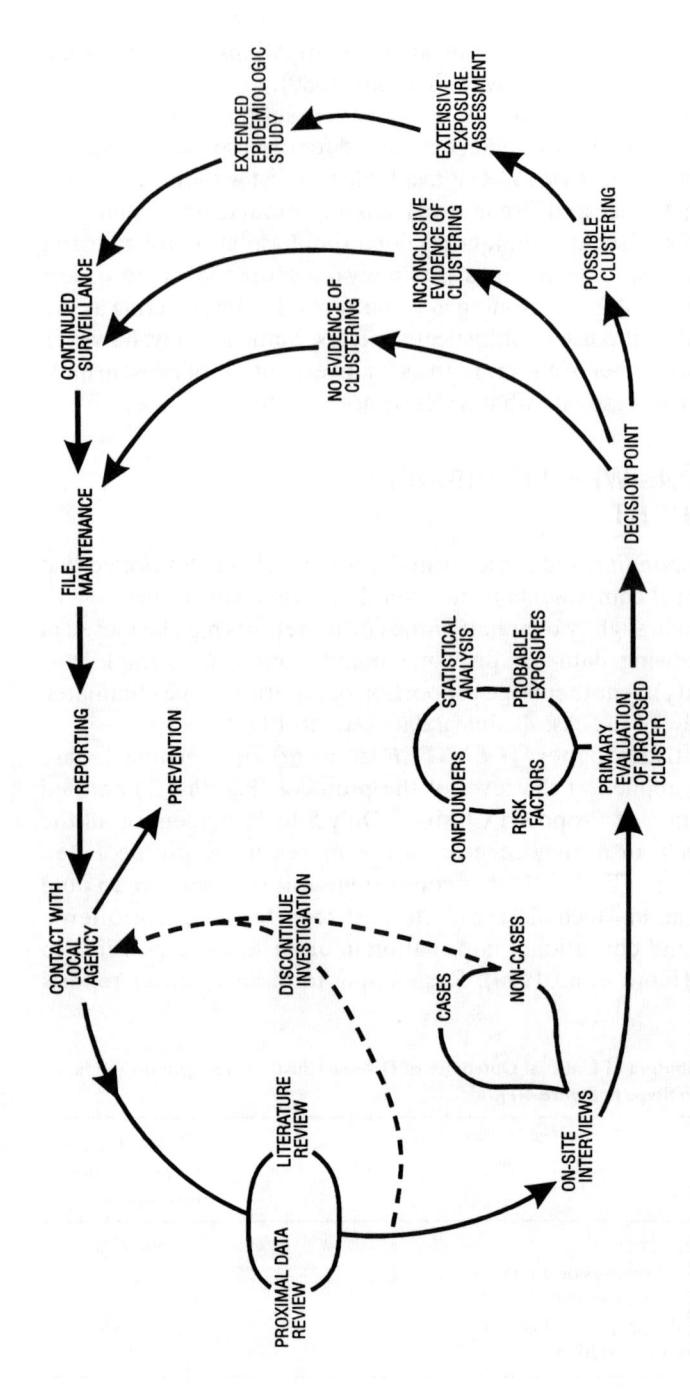

FIGURE 4-1. Flow Chart for Responding to Reports of Space-Time Disease Clusters.(Refer to Table 4-2 for Additional Clarification).

involves more than a search for disease etiology, for after all, health professionals must also consider mental and social well being (Bender et al. 1990). The following questions are provided as guidance in considering evidence of disease clustering:

1. The first question is always, "Is this cluster a random phenomena?"

This question has no simple answer. How do we declare what is expected? How do we measure randomness? We need some parameters, and there are several classic parameters: space, time, series, and combinations of events. The statistical methods described later in this chapter, and especially the techniques in the *CLUSTER* software package, seek to answer specific questions phrased around one or more of these parameters.

2. Are these cases more closely grouped in space than might result from random placement?
3. Do these cases group closer in time than would occur by random distribution?
4. Do these cases aggregate closer in space and time than might be expected from a random pattern of occurrence?

These are the classic questions. For answers, we turn to statistical methods that require data for each case related to its location in space, time, or both space and time. The methods in the *CLUSTER* software will evaluate cluster data in several different ways, and these are the essential questions addressed.

5. Do some geographic units (for example, communities, counties, zipcodes) contain more cases in them than might be expected from random incidence?
6. Do some population groups (for example, communities, ethnic groups) experience more cases than their proportion of the population might indicate would be expected from random incidence?

These variations on the space-time questions involve patterns of cells (for example, adjacencies of cells having similar disease experience), or they rely on conventional occupancy logic (for example, the simple dependent probabilities of placing a number of cases in some number of cells randomly). These cell-occupancy methods are more complex than the straightforward space-time approaches, and they have special data requirements. We should know in advance how the data should be organized, and the kinds of questions that will need to be answered with these techniques.

7. Does the pattern of case occurrence indicate that over time a change in disease frequency has occurred?

8. Do the observed rates of disease suggest that a statistically significant departure from the expected rates has occurred recently?

Remember the fundamental validity and precision errors discussed in Chapter 3 (for example, biases, multiple comparisons, etcetera), these must be considered with cluster studies as well. Complex statistical techniques do not overcome these basic epidemiologic considerations. In fact, **cluster statistics are excruciatingly susceptible to these influences**.

9. How do we interpret statistical significance?

There are two important considerations to take into account with regard to statistical significance related to disease cluster analyses:

   a. Statistical techniques for studying disease clusters were designed for use with rare events (for example, those with incidence rates < 10/100,000); most of them may be performed with as few as three cases. This is unusual for statistics: be cautious! Ask a few questions of your own: Are the observed cases consistent with any suspicions that you have about the pattern of disease occurrence? Does this pattern fit with some other pattern that you are aware of within the community (for example, a foci of high-risk persons)? Does more than one statistical test suggest evidence of clustering?
   b. The statistics presented with ***CLUSTER*** software do not all have conventional or normal test statistics like the z-test. As a rule they mostly use the Poisson distribution (for example, useful for instances where only one or fractions of cases are expected). Therefore, be a bit more rigorous with interpretation than you might usually be (for example, $p < 0.05$ may not be good enough to take public health action with a cluster study). Use your experience and judgment: If several statistical methods all show marginal evidence of clustering ($p < 0.05$), and you have other reasons to be suspicious of the pattern (the cluster is near a recognized environmental hazard), take positive action, even in the face of usual statistical criteria. The space-time methods (questions 2–4), and the cell-occupancy methods (questions 5–6) can also be used for surveillance studies (questions 7–8) with some modifying assumptions. There will be additional information regarding surveillance methods in Chapter 5.

THE CLUSTER SOFTWARE

CLUSTER presents a strategy for **screening** reports of disease clusters that is based on a battery of statistical tests. These tests are used to identify clusters that warrant further epidemiologic evaluation (Fiore et al. 1990; Editorial

1990). This screening battery for disease clusters provides a valuable middle ground between the extremes of investigating all cluster reports (Langmuir 1965; Caldwell and Heath 1976), and not investigating cluster reports at all (Rothman 1990; Neutra 1990).

Reports of disease clusters plague public health professionals, notably reports of cancer clusters and their relationship to potential environmental contamination from hazardous point sources (Rothman 1990; Thacker 1989). A recurring theme among the reports of disease cluster investigations is that very few (approximately 5 percent are productive research endeavors based on cost/benefit considerations) (Guidelines 1990; Schulte et al. 1987; Garfinkel 1987; Bender et al. 1990). A persistent policy question has been how to sort through the many cluster reports that an agency receives in order to arrive at a productive subset of reports with minimal effort and cost.

Many states have developed protocols for responding to disease cluster reports that seek to rank order them for improved productivity (Warner and Aldrich 1988). Recent work at the United States Centers for Disease Control (CDC) has focused on developing a standard protocol for evaluating disease cluster reports (CDC 1990). The CDC has also conducted research toward more subtle means of recognizing trends in temporal data for monitoring reportable disease (for example, the basis for when an **epidemic** increase is declared) (Stroup et al. 1989).

Perhaps the best-known statistical test for detecting disease clustering is that of Knox (1964). Knox's method uses a contingency table approach where case-pairs are assigned to cells of a 2-by-2 table on the basis of their satisfying either, neither, or both user-defined criteria for **closeness** in space and time (Figure 4-2). This straightforward test for space-time interaction is the cornerstone for the battery of statistical tests proposed for evaluating reports of clusters of rare health events (usually less than 30 cases) (Heath 1990). Since the design of the Knox approach is so simple, it may be used with lectures in the community to explain the objectives of the statistical, cluster analyses.

It is usually a good strategy to use more than one statistical approach to assess evidence of unusual disease aggregation. The computer program *CLUSTER* contains a battery of 12 statistical tests: Two tests each for space-time interaction, temporal clustering, cell-occupancy, geographic (nearest neighbor) clustering, time-series analysis, and birth defects monitoring (Aldrich and Drane 1990). Methods within a single category should generally agree; consistency between findings reinforces results that may be of borderline significance ($p < 0.05$). With multiple statistical methods being used, agreement by three or more methods may be regarded as remarkable. If five or more methods agree, this is cause for overwhelming attention to the clustering event.

Remember, however, that statistical methods only assist causal reasoning; consistency, biologic plausibility, and dose-response relationships should always outweigh the results of a statistical test. Yet p-values can be very helpful when

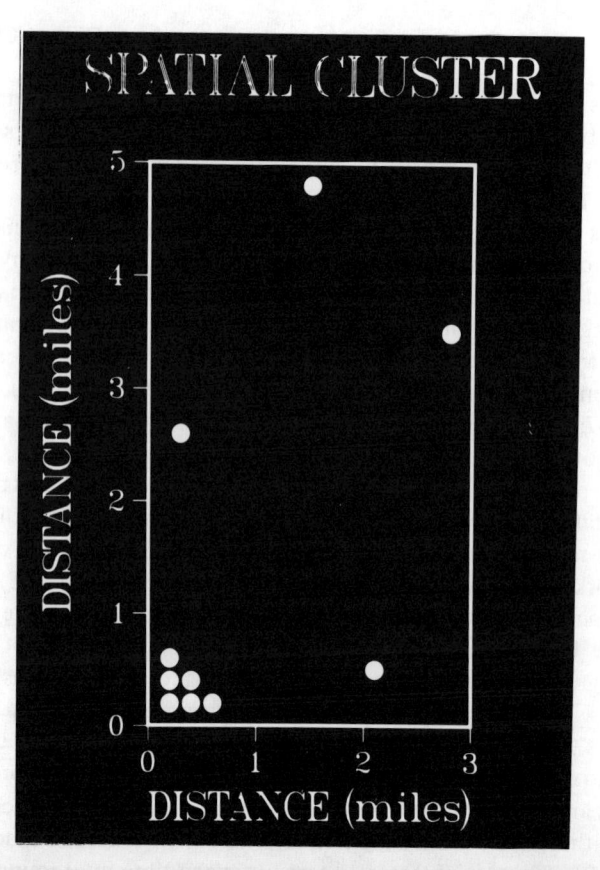

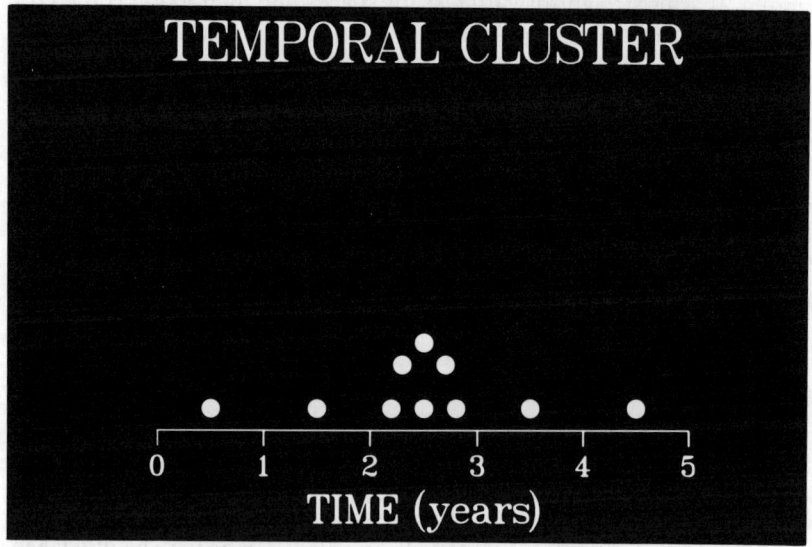

FIGURE 4-2. Digrammatic Examples of Closeness in Space, Time, and Space-Time Interaction vis-a-vis Knox Method for Detecting Space-Time Interaction.

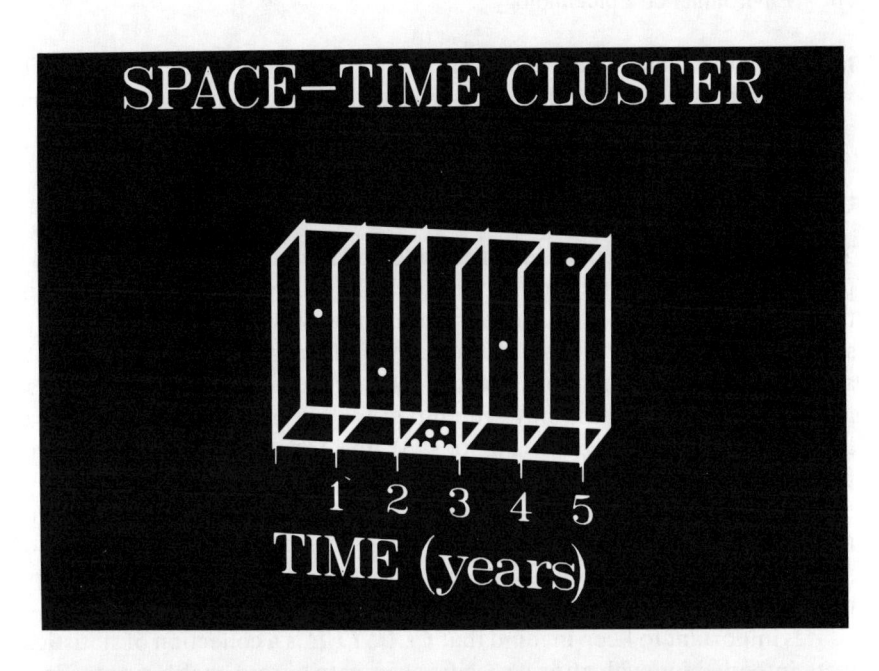

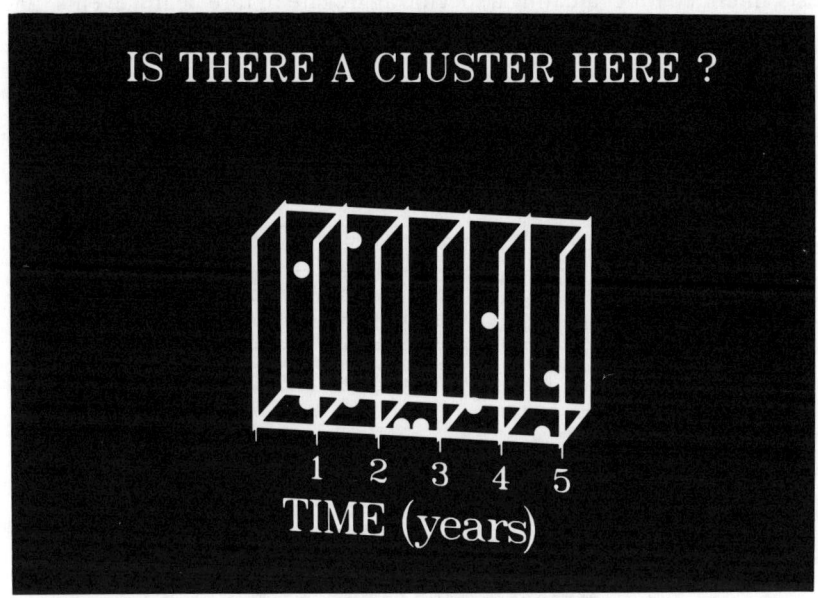

FIGURE 4-2. *(continued)*

70   Environmental Epidemiology

faced with the need to make decisions in the face of uncertainty (for example, when one must select which cluster reports deserve further study) (Fiore et al. 1990).

Using the *CLUSTER* Computer Program

This section provides the user of the *CLUSTER* software package with a perspective on the relationships between the various statistical tests contained in the software. By understanding the way the methods evaluate similar characteristics, the user may elect to perform several tests to complete an analysis. All of the methods in *CLUSTER* are tests for evidence of disease aggregation, and as such, they are unified by that hypothesis. However, as the opening menu suggests (Figure 4-3), the tests use very different approaches. These differences are important because a variety of approaches to the same problem can add or subtract information from findings that are marginal or on the edge of statistical significance. With the strategy of statistical tests as a battery (using multiple methods), *CLUSTER* offers the user multiple results for evaluating a suspected cluster.

It is important to keep in mind that *CLUSTER* is a collection of statistical tests, intended for evaluating a specific aggregate of rare health events. Each

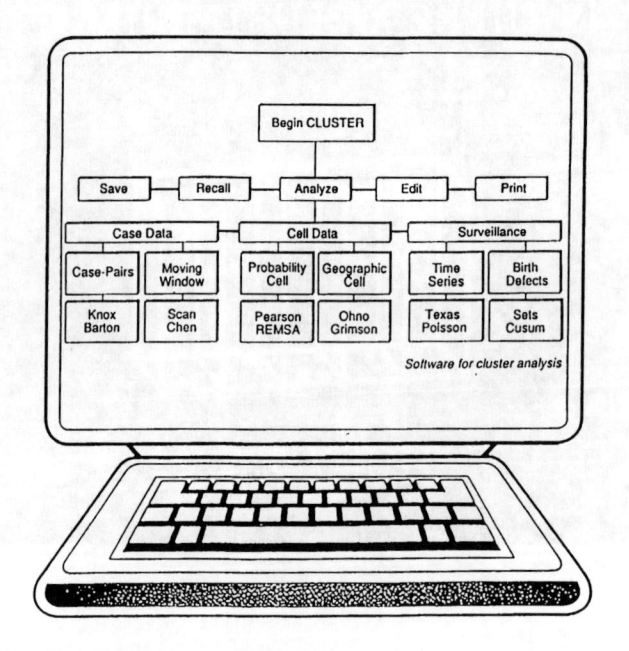

FIGURE 4-3.   Opening Screen of *CLUSTER* Software.

TABLE 4-2. Anticipated Agreement of Methods for Similar Disease Aggregates

	Knox	Barton	Scan	Chen	Grimson	Ohno	Pearson	REMSA	Poisson	Texas	Sets	CuSum
Knox	—	XXX					XXX	XXX				
Barton		—					XXX	XXX				
Scan			—	XXX					XXX	XXX		XXX
Chen				—	—						XXX	
Grimson					—	XXX						
Ohno						—						
Pearson							—	XXX				
REMSA								—				
Poisson									—	XXX		XXX
Texas										—		XXX
Sets											—	XXX
CuSum												—

of the methods in the six major data entry paths are related (for example, continuous data, geo-cell, time series). Each of these groups of methods should generally agree (Table 4-2). It is a good strategy to use all of the methods in a particular data entry path, and then inspect each of the results for consistency. Consistency would reinforce statistical findings that were on the edge of being deemed significant (for example, $p < 0.05$).

The following is a general guide to the methods by data entry paths:

Case-pairs: Knox is quite sensitive, and it will find more significant event aggregates than most methods (Knox 1964). Barton (1965) is very likely the best method in CLUSTER, but is less sensitive than Knox. Alone, it is a strong case for clustering; in agreement with either of the other case-data methods, and it is a basis for confidence in a finding of clustering. Because Knox uses a threshold, the test does not use all of the available data. The Barton method uses all available data and is reliable for circumstances of spatial or temporal increases.

Moving window: These are temporal methods only (no analysis of space is involved), and therefore will likely not agree with the continuous methods. Chen is the best of the methods in this section, but as developed by the authors, it requires all of the cases to have reduced intracase intervals to declare a cluster (Chen et al. 1982). This criterion is a bit stringent and makes the method less useful than it might be in instances where several case-pairs have shortened intercase time intervals (the decision criterion recommended with **CLUSTER** is that a majority, or five or more of the cases, have shortened intervals). The scan test is the most versatile moving-window test and the most sensitive (Wallenstein 1980; Naus 1982). It may not always agree with Chen, but often it will be consistent with Barton.

Probability cells: This is one of the more innovative strategies for considering evidence of clustering. These methods evaluate whether the observed pattern of disease occurrence is random. To do this, the methods assume randomness, unless some sort of **force** is operating that causes them to be aggregated in some manner. This force is assumed to be evidenced by the amount of total available space occupied by a small number of the cases.

The smaller the space occupied by the larger number of cases, the further the departure from randomness. This is what is meant by *probability cells:* The solutions offered by the two methods are variations on the theme of evaluating randomness. Pearson's method uses a chi-square goodness-of-fit test (Pearson 1912; Drane and Hua 1982). It is particularly good for looking at city blocks, households, and other symmetric areas. REMSA is the **inverse** of Pearson; it starts with the cluster and asks about the remaining available space (Aldrich et al. 1985; Parzen 1960). REMSA may be used with a variety of variables: age, race, sex, as well as space and time. Together, these methods should offer considerable agreement. REMSA and Pearson should agree with the space-time tests (Knox, Barton).

Geo-cells: Data entry for these methods can be cumbersome with many cells. These methods are designed to evaluate the proximity of cells with similar disease experience. The method by Ohno et al., (1979) uses a chi-square solution and may be used with several levels of high occurrence. With Grimson et al. (1981) a hierarchial solution is derived for the disease rate defined as "high." These methods are best applied to large areas (for example, states). Municipalities may also be studied using census tracts or zip codes. These two methods should agree with one another to the extent that the clusters identified should be intercepts of one another. As these methods are cross-sectional (for example, no time perspective), they should be compared with any of the other methods only from a spatial perspective.

Time Series: As data accumulates over time, these two methods test the sequential intervals for evidence of a departure from expected patterns. The methods test this time trend against two solutions: the Poisson distribution and the negative binomial (Grimson 1979). With the Texas method (Hardy et al. 1990), a decision criterion must be set for declaring an increase; this innovative two-step decision technique builds in a protective factor for a false-positive finding. With the Poisson method (Hill et al. 1968), an alternative solution using the negative binomial may be selected for verifying a consistent pattern of increase. Generally, the time series will be more sensitive to temporal aggregates than the moving-window techniques, yet the latter should declare a maxima in the same area along the time line as the increase detected by the time series.

Birth defects: These methods are for use with studies of birth defects. Sets searches for shortening the intercase interval as evidence of an increasing rate (Chen 1978). With CuSum, the objective is to detect a run of defects greater than expected based on the rate of live births (Chen et al. 1983). It is pivotal that there be uniformity of the time intervals associated with case occurrence (for example, the period of gestation). For events that are expected to have varying time intervals (for example, cancers), these methods may not be appropriate. These two methods are so different there is a good chance they will not agree with one another. However, some of the other methods may be useful with birth defects. The Chen test could be expected to agree with the Sets method since it uses the same strategy. The Scan and Texas methods should agree with the CuSum test.

OTHER METHODS

The statistical methods in the ***CLUSTER*** software are not all that have been developed for evaluating disease aggregates. Several other fine methods were candidates for ***CLUSTER*** including:

Empty cells by Grimson (1992): This is a test for **runs** of zeros in rare event occurrence data.
Mantel regression technique (Mantel 1967): This is a continuous version of Knox, and works well with larger clusters (for example, > 20 cases).
Ederer, Myers, and Mantel (1965; Ederer et al. 1966; Mantel et al. 1976): This

cell-occupancy approach tests the probability of the maximum observed aggregate within cells, being a random phenomenon.

Pike and Smith (Pike and Smith 1968; Pike and Smith 1974). This is an adjacency technique, like the geo-cells, yet it uses case-pair data and can be very complex to perform.

Spatial Autocorrelation (Glick 1979). This is a parametric method for evaluating case aggregates and has been frequently modified. The primary detraction of this method is that it is very computation-intensive. Some investigators have also used a simple contingency table to evaluate whether cases lived closer to a point source than comparison subjects (Matanowski et al. 1981; Lyon et al. 1981). Surveillance study methods are discussed in Chapter 5.

STRATEGIES FOR USING STATISTICS TO TEST FOR DISEASE CLUSTERING

In this section we discuss some of the strategies associated with performing statistical analyses of disease cluster data. Although we will not discuss all of the strategies, those we discuss will be so sufficiently mainstream that a review is worthwhile, especially in light of the advent of the ***CLUSTER*** software.

When one attempts to evaluate apparent increases in the occurrence of more common disease endpoints, one must consider confounders. Thus the wide use of the conventional case-control method (see Chapter 3 for a more detailed discussion of the case-control design). While the case-control design is more analytic, the cluster analysis is more exploratory. When evidence of disease clustering is strong, a case-control study may be used to test hypotheses suggested from the cluster analysis.

As a matter of analysis, even small numbers of cases may lead to large calculations because of the many methods that use case-pair data. For example, 10 cases will make 45 possible combinations of case pairs ($[n]$ * $[n-1]/2$). With geographic or probability cells, the number of cells is an independent consideration to the scale of an analysis. Often the empty cells are simply ignored, but sometimes that may not be the case. For example, with studies over time, even a very sparse pattern of cases will lead to many scans of a time interval. A cluster report that involves small geographic areas and short time periods will ease analyses.

A troublesome aspect of statistical methods for evaluating clustering is that many assume a uniform distribution in the underlying population. However, some are not so dependent on this uniformity (for example, the methods that use case-pair data). Several methods are quite susceptible to simply finding a nonuniformity of the population (for example, a densely populated residential area), rather than a meaningful cluster. To adjust for these realities, two strategies are usually used:

1. With the geographic cells, population characteristics of age, race, sex adjusted rates are used, rather than case counts or crude rates; and
2. With the moving-window analyses the width of the time intervals may be weighted for growth of the population.

This is done in a posterior fashion described more fully with the scan test literature (Wallenstein 1980). However, in any of these instances, such adjustments for the underlying population make the analyses more complex and usually require processing the data before analysis.

Recall that the strategy implicit with all of these methods is to detect a clustering of cases: That goal is not always the same as detecting a generalized increase in disease rates. Some of the methods included were developed for a surveillance application (for example, goodness-of-fit type methods); that is, expressly to detect an increase in disease rates over time. These techniques assume a consistency of the background disease rates, and this assumption is analogous to the uniform population assumption with spatial analyses. We should keep these implicit strategies in mind when selecting clustering meth-

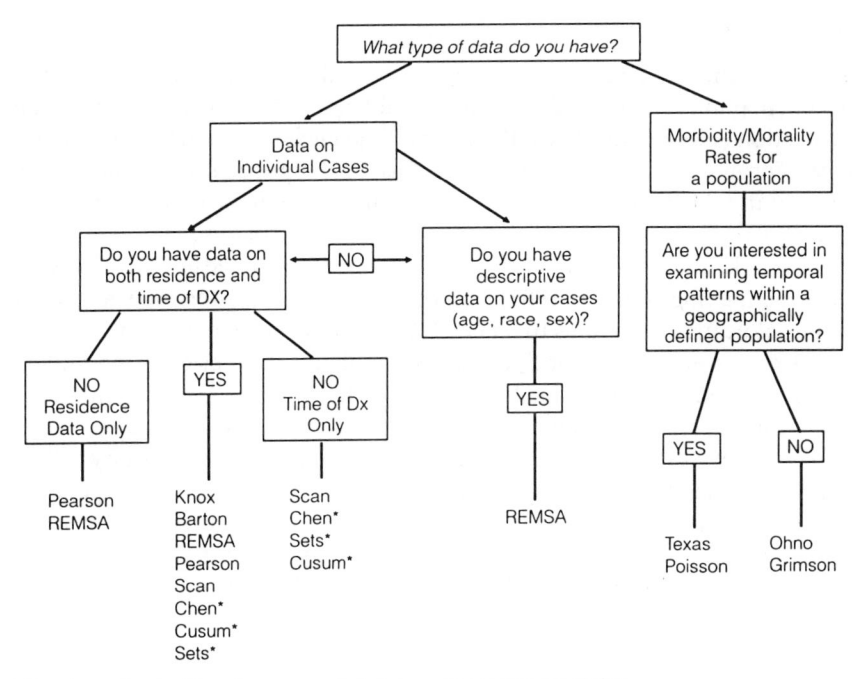

* Requires estimate of baseline or expected disease rates in your population.

FIGURE 4-4. Flow Chart for Selecting Methods for *CLUSTER*.

ods (Figure 4-4). This is exactly why analyzing a set of data with several cluster methods is such a sound strategy. Different methods will have different inherent weaknesses; when several techniques reinforce one another they do so by compensating for strengths and fallacies.

Most cluster reports relate to past occurrence of disease, cases that have already occurred. Surveillance, by contrast, has as its goal the detection of cases (clusters) as they occur so that intervention activities can be applied. Several of the methods in the ***CLUSTER*** software package are especially useful for surveillance application (CuSum, Texas, Poisson, REMSA, Chen). With the available methods, you should feel free to use a combination of methods approach. That is, you should use the methods especially intended for surveillance studies to identify a candidate aggregation, then perform simulations with the other (retrospective) methods to reinforce space-time characteristics of the cluster. **Cluster analyses are exploratory methods.** Their greatest value is not in attempting to prove or disprove a cluster report, but in assisting the epidemiologist to make a causal interpretation by refining judgment and distinguishing between difficult choices.

SUMMARY

The evaluation of disease cluster reports would seem to be as permanent a fixture in public health practice as is the public's concern over environmental protection (Fiore et al. 1990). Public health agencies are in need of technology to promote general welfare in the face of the public's apparent wish to have both a highly industrialized society and a safe environment (Ruckelshaus 1984). Tests to detect disease clustering, while not designed to estimate disease risk, may be used to alert agencies to potential need and to assist with decision making (Glasser 1985). Reports for disease clusters may also alert public education programs to a need for information dissemination.

As discussed in Chapter 5, disease surveillance is an established and an important part of public health practice (Thacker and Berkelman 1988). Increasingly, there are available population-based resources for public health surveillance (for example, 41 states have cancer registries)(Utility 1990). With the recurring and persistent problem of disease cluster reports to consider, a surveillance strategy is recommended that would provide an active approach to disease clustering: This strategy is the systematic monitoring of disease patterns expressly to detect unusual case aggregations before they present themselves as a cluster **report.** For this purpose, we recommend the selection of strategic, rare health events as sentinels of potential exposure to hazardous substances in the environment (Rothwell et al. 1991). Rare health events may serve public health surveillance particularly well because active responses to reports of small area disease aggregates (clusters) may prove a useful means for encouraging dialogue between concerned citizens and agency representatives (Fiore et al. 1990). However, systematic processes must be carefully organized around a consistent protocol and should be guided by statistical methods applied as a screen for selecting promising reports for further studies.

Epidemiology should move forward with developing new and more efficient cluster identification methods for use with population-based disease surveillance of environmental health concerns (Aldrich et al. 1989; Monson 1990).

ASSIGNMENTS

Consider the following data from an imaginary cluster in Hedrack County between 1985 and 1989. Ten acute myelocytic leukemia cases were diagnosed among this community's residents. The residents were distressed because of a large fire at a local wood treating plant. The diagrams in Figure 4-2 may be used to visualize these cases. Assume that the community (represented by the rectangle is uniformly distributed within the space; and that the wood treating plant is at the 0,0 point.)

Cases	Latitude	Longitude	Date of Diagnosis	Age	Race	Sex
1	28	75	6/1985	40	W	M
2	5	35	2/1986	65	B	F
3	8	6	1/1987	10	B	F
4	3	3	3/1987	12	B	M
5	7	2	4/1987	7	B	M
6	1	2	8/1987	8	B	M
7	9	1	9/1987	5	B	F
8	5	5	12/1987	7	B	M
9	31	9	7/1988	58	B	M
10	44	54	10/1989	71	W	F

1. Use the ***CLUSTER*** program (if you have it available) to run the Knox analysis with these data. Use threshold for "close" of 12 space units and 12 months. If you don't have the ***CLUSTER*** program, the cells in the contingency table are: A = 15, B = 0, C = 5 and D = 25. You test the "A" cell for significance. You can do this by hand in much the same way you would solve for a chi-squared solution (for example, $(A + B) \times (A + C)/$ total pairs (45 here). Be sure that your testing of the A cell probability uses a Poisson probability.
2. These results will be quite statistically significant, yet if the time interval were not so sharp (for example, if there were two cases in each year, the results from Knox drops to three case-pairs are "close" and four are

expected, p < 0.709). Discuss the implications with these time values from a "real world" cluster involving cancer (with variable latency).
3. Use the program Scan with these data using one time period for the window. If you do not have **CLUSTER** software, look at the time series of cases as a run: 1,1,6,1,1. Does it surprise you that two cases would be expected in each time period? With two cases expected, six observed in one cell is p < 0.117. Would you still investigate these cases further?
4. The previous two methods (Knox and Scan) are tests for space-time interaction and for temporal clustering respectively. The REMSA test uses person characteristics as well as place and time. If you have **CLUSTER** software, perform the analysis with these five parameters, with these number of classes and population proportions:

Parameter	Number of Classes	Percentage of Population In This Cell	Explanation
Age	8	0.15	(Children under 10 yr)
Race	2	0.25	Blacks only
Sex	2	0.5	Male
Year	5	1987	Year of interest
Quandrant	4	Lower left	Where the cluster is

This will give you 640 cells; the cell of interest (black boys under age 10 with acute myelogenous leukemia in 1987 who live in the lower right hand quadrant of the community) has a random probability of 0.000938. What is the probability of four of the ten cases occurring with these person, place, and time characteristics? Figure 5-7 contains the solution formula if you don't have **CLUSTER**. How do you respond to this result? What are your thoughts about the differing results from these three tests?

GLOSSARY

Cells: There are two types of cells referred to in cluster analysis: **geographic cells** and **probability cells.** The approach to the analysis of both of these cells differs markedly. Geographic cells are areas defined for other than epidemiologic purposes (for example, counties, postal districts, voter precincts). In the analysis of geographic cells, the conventional approach is to examine the pattern of adjacency (that is, Do cells with similar disease rates cluster?). Although a correlation technique does exist, it analyzes geographic cell data as continuous variables, and it is very complex! Probability cells are arbitrarily defined cells that are often just a variation on spatial analyses (that is, cells are defined as grids of the X, Y coordinates). These cells may be used as demographic probabilities too (that is, age, race, sex, etcetera). The same general assumptions apply as with spatial analyses, but rather than calculations based on case-specific data, or case-pair data, the analysis is for the pattern of occupancy by the cases.

Cluster: An aggregation of independent events that is deemed to be a departure from expectation (that is, not a random or chance phenomena).

Goodness-of-Fit: With disease surveillance, some clustering methods are simple comparisons of the observed experience to an expected distribution of cases. However, these methods often have subtle compensations for repeated measures and for avoiding Alpha (Type-I) errors (that is, TEXAS, Chen, Sets). These methods may also employ the two-step decision process (that is, alert and action levels of significance).

Space: The place where the case occurred. Usually a place of residence is used, or a place of employment. Space is defined in linear coordinants, the units are arbitrary: miles, blocks, kilometers, etcetera. Think of distance as so many units. Usually space is considered as X,Y values on a grid surface with a fixed 0,0 point. A key underlying assumption to most spatial analyses is that the surface is uniform (that is, the underlying population is not grouped somehow and all subjects share equal risk). **NOTE: All spatial programs expect X,Y coordinates. So overlaying a grid onto your study area will be useful.**

Time: The time when the case occurred. Units are arbitrary: months, weeks, or years. Months are the most common increments, yet years, days, and weeks have also been used. As with space, the study population is presumed to be stable and uniform over time. **NOTE: Dates are entered as "M M D D Y Y" in CLUSTER and MUST be in sequential order.**

REFERENCES

Aldrich, T. E. and Drane, J. W. 1990. *CLUSTER: User's Manual for Software to Assist with Investigations of Rare Health Events.* Published by the Agency for Toxic Substances and Disease Registries, Atlanta, GA.

Aldrich, T. E., Atkinson, D. A., Hines, A., Smith, C.G. 1990. "The Establishment of a Population-Based Cancer Registry for North Carolina." *North Carolina Medical Journal* 51(2), pp. 107–112.

Aldrich, T. E., Meyer, R. E., Qualters, J., and Atkinson, D.A. 1989. "Rare Health Events as Sentinels of Environmental Contamination." In *Proc. Public Health Conference on Records and Statistics*, DDHS Pub. No. 90-1214: 323–6.

Aldrich, T. E., Wilson, C. C., and Easterly, C.E. 1985. *Population Surveillance for Rare Health Events.* Proc. Public Health Conf. on Records and Stat., DHHS Pub. No. (PHS) 86-1214:215–220.

Barton, D. E., David, F. N., and Merrington, M. 1965. "A Criterion for Testing Contagion in Time and Space." *Annals of Human Genetics* 29:97–101.

Bender, A. P., Williams, A. N., Johnson, R. A., and Jagger, H.G. 1990. "Appropriate Public Health Responses to Clusters: The Art of Being Responsibly Responsible." *American Journal of Epidemiology* (132):S48–S52.

Caldwell, G. G. 1989. "Time-Space Cancer Clusters." *Health and Environment Digest* 3(5):1–3.

Caldwell, G. G., and Heath, C. W. 1976. "Case Clusters in Cancer." *Southern Medical Joournal* 69:1598–1602.
CDC 1990. *Guidelines for Investigating Clusters of Health Events, Morbidity and Mortality Weekly Report Vol. 39* (RR-11), 23 pages, 35 references. Available from the Mass. Med. Soc., C.S.P.O. Box 9120, Waltham, MA 02254-9120 for $3.00.
Chen, R. 1978. "A Surveillance System of Congenital Malformations." *J. American Statistical Association* 73:323–27.
Chen, R., Mantel, N., and Isaacson, C.P. 1982. "A Monitoring System for Chronic Diseases." *Meth. Inform. Med.* 21:86–90.
Chen, R., McDowell, M., Teraian, E., and Watherall, J. 1983. *Eurocat Guide to Monitoring Methods for Malformation Registers*. EEC Concerted Action Project Bruxelles.
Doll, R. 1981. "Relevance of Epidemiology to Policies for the Prevention of Cancer." *J. Occupational Medicine* 23:601–9.
Drane, J. W., and Hua, T. A. 1982. "Decomposing Three Dimensional Contingency Tables." *Statistische Hefte*. 23:122–27.
Ederer, F., Myers, M. H., Eisenberg, H., and Campbell, P.L. 1965. "Temporal-Spatial Distribution of Leukemia and Lymphoma in Connecticut." *Journal of the National Cancer Institute* 35:625–29.
Ederer, F., Myers, M. H., and Mantel, N. 1966. "A Statistical Problem in Space and Time: Do Leukemia Cases Come in Clusters?" *Biometrics* 20:626–28.
Editorial. 1990. "Disease Clustering: Hide or Seek?" *Lancet* 336:717–18.
Fiore, B. J., Haranhan, L. P., and Anderson, H. A. 1990. "State Health Department Response to Disease Cluster Reports: A Protocol for Investigation." *American Journal of Epidemiology* 132:S14–S22.
Garfinkel, L. 1987. "Cancer Clusters." *CA: A Cancer Journal for Clinicians* (37):20–25.
Glasser, J. H. 1985. *Health Statistics Surveillance Systems for Hazardous Substance Disposal*. Proceedings from the 1985 Public Health Conference on Records and Statistics, DHHD Pub. No. (PHS) 86-1214:221–224.
Glick, B. 1979. "The Spatial Autocorrelation of Cancer Mortality." *Social Science and Medicine* 13D:123–30.
Grimson, R. C. 1979. "The Clustering of Disease." *Mathematical Biosciences* 46:257–78.
Grimson, R. C., Wang, K. C., and Johnson, P. W. C. 1981. "Searching for Hierarachial Clusters of Disease: Spatial Patterns of Sudden Infant Death Syndrom." *Social Science and Medicine* 15:287–93.
Grimson, R. C.,Aldrich, T. E., Drane, J. W. "1992 Clustering in Sparse Data and an Analysis of Rhabdomyosarcoma Incidence." *Statistics in Medicine* 11:761-768.
Guidelines for Investigating Clusters of Health Events. 1990. *Morbidity and Mortality Weekly Report* 39:1–23.
Hardy, R. J., Schroeder, G. D., Cooper, S. P., Buffler, P. A., Prichard H. M., and Crane, M. 1990. "A Surveillance System for Assessing Health Effects from Hazardous Exposures." *American Journal of Epidemiology* 132:S?–S?.
Heath, C. W. 1990. "Author's Reply to Thymic Function and Leukemia." *CA: A Journal for Cancer Clinicians* (40):319–20.
Hill, G. B., Spicer, C. C., and Weatherall, J. A. C. 1968. "The Computer Surveillance for Congenital Malformations." *British Medical Bulletin* 24: 215–18.

Houk, V. N, and Thacker, S. B. 1987. "Registries: One Way to Assess Environmental Hazards." *Health and Environment Digest* 1:5–6.

Knox, E. G. 1964. "The Detection of Space-Time Interactions." *Applied Statistics* 13: 25–29.

Langmuir, A. D. 1965. "Formal Discussion of the Epidemiology of Cancer: Spatial and Temporal Aggregation." *Cancer Research* (25):1384–86.

Lyon, J. L., Klauber, M. R., Groff, W., and Chiu, G. 1981. "Cancer Clustering Around Point Sources of Pollution: Assessment by a Case-Control Methodology." *Environmental Research* 25: 63–74.

Mantel, N. 1967. "The Detection of Disease Clustering and A Generalized Regression Approach." *Cancer Research* 27:209–20.

Mantel, N., Kryscio, R. J., and Myers, M. H. 1976. "Tables and Formulas for Extended Use of the Ederer-Myers-Mantel Disease Clustering Procedure." *American Journal of Epidemiology* 104:576–84.

Matanowski, G. M., Landau, E., Tonascia, J., Lazar, C., Elliott, E. A., McEnroe, W., and King, K. 1981. "Cancer Mortality in an Industrial Area of Baltimore." *Environmental Research* 25:8–28.

Monson, R. R. 1990. "Editorial Comment: Epidemiology and Exposure to Electromagnetic Fields." *American Journal of Epidemiology* (131):774–75.

Naus, J. I. 1982. "Approximations for Distributions of Scan Statistics." *J. American Statistical Association* 77:177–83.

Neutra, R. R. 1990. "Counterpoint from a Cluster Buster." *American Journal of Epidemiology* (132):1–8.

Ohno, Y., Aoki, K., and Aoki, N. 1979. "A Test of Significance for Geographic Clusters of Disease." *International Journal of Epidemiology* 8:273–81.

Parzen, E. 1960. *Modern Probability Theory and Its Application.* New York: John Wiley and Sons, Inc.

Pearson, K. 1912. "On the Appearance of Multiple Cases of Disease in the Same House." *Biometrika* 8: 404–35.

Pike, M. C. and Smith, P. G. 1974. "A Case-Control Approach to Examine Disease for Evidence of Contagion, Including Disease with Long Latency Periods." *Biometrics* 30:263–79.

Pike, M. C., and Smith, P. G. 1968. "Disease Clustering: A Generalization of Knox's Approach to the Detection of Space-Time Interactions." *Biometrics* 24: 541–56

Rothman, K. J. 1990. "A Sobering Start to the Cluster Busters' Conference." *American Journal of Epidemiology* (132):S6–S13.

Rothwell, C., Leaverton, P., and Hamilton, C. 1991. "Identification of Sentinel Health Events As Indicators of Environmental Contamination: Report of a Consensus Development Conference." *Environmental Health Perspectives* 94:261–63.

Ruckelshaus, W. D. 1984. "Risk In A Free Society." *Risk Analysis* (4):157–62.

Schulte, P. A., Eherenberg, R. L., and Singall, M. 1987. "Investigation of Occupational Cancer Clusters: Theory and Practice." *American Journal of Public Health* (77): 52–56.

Stroup, D. F., Williamson, G. D., and Hearndon, J. L. 1989. "Detection of Aberrations in the Occurrence of Notifiable Disease Surveillance Data." *Statistics in Medicine* (8):323–29.

Thacker, S. B. 1989. "Time-Space Clusters: The Public Health Dilemma." *Health and Environment Digest* 3(5):4–5.

Thacker, S. B., and Berkelman, R. L. 1988. "Public Health Surveillance in the United States." *Epidemiologic Reviews* (10):164–90.

Utility of Cancer Registries Varies. 1990. *Public Health Macroview Vol. 3* (5):2–3.

Wallenstein, S. 1980. "A test for Detection of Clustering Over Time." *American Journal of Epidemiology* 11:367–72.

Warner, S. S., and Aldrich, T. E. 1988. "The Status of Cancer Cluster Investigations Undertaken by State Health Departments and the Development of a Standard Approach." *Journal of the American Public Health Association* 78(3):306–7.

5

Surveillance Activities in Disease and Exposure Situations

Tim E. Aldrich and Jack Griffith

OBJECTIVES

This chapter will:

1. Review criteria for the development of a surveillance program.
2. Identify the components of a surveillance program.
3. Discuss analytic issues in surveillance systems.
4. Describe innovative modifications for environmental epidemiologic applications.

BACKGROUND

The **continuing watchfulness** of all aspects of the occurrence and spread of disease that is pertinent to effective control measures has been defined as surveillance (Thacker and Berkelman 1988; Mausner and Bahn 1974).

Public health scientists measure the hazard of environmental health problems, as well as the effectiveness of remedial programs, with human exposure and disease data. Unfortunately, **real world** data on environmental exposures, and human disease data are very often fragmented, incomplete, inappropriate, or entirely lacking. The absence of adequate exposure and disease data leaves risk assessors and risk managers with no choice but to rely on theoretical estimates of exposures and disease. Thus, estimates may not be representative of the actual human experience.

The development of nationwide disease surveillance activities date back to the early 1950s when the CDC of the United States Public Health Service began to cooperate with state health departments to collect information on malaria cases. Surveillance activities grew through the years, until we now have nationwide reporting of hepatitis, influenza, tuberculosis, venereal disease (including the HIV virus), food-borne disease, vehicular accidents, and hospital acquired (nosocomial) infections.

Today, surveillance techniques range from gathering and disseminating routine morbidity and mortality data to the analysis and dissemination of data from highly sophisticated laboratory and field investigations.

DEVELOPING A SURVEILLANCE PROGRAM

Essential Components of an Effective Surveillance Program

Surveillance activities include case finding, exposure assessment, epidemiologic research, and the dissemination of epidemiologic and health data for managing public health problems (for example, risk assessment calculations, the evaluation of the effectiveness of public health programs, and the impact of regulatory activities by responsible government authorities). Essential components of an effective and ongoing surveillance program include the **collection, collation, and analysis** of relevant data, followed by appropriate and regular reports to responsible authorities.

As shown in Figure 5-1, several factors to be considered in the development of a surveillance activity are:

1. **Case finding involves providing information on disease sources, identification of cases, and measurement of the problem within a given community. Case finding** is based on the disease outcome of interest and the potential exposures under study. Although there are several classification systems, the **International Classification for Diseases (ICD)** is probably the most familiar. The ICD is used by all hospitals and insurance systems. A reportable list of diseases must be developed for the cases to be identified. Next a system must be put into place to identify the occurrence (i.e. ascertainment) of these cases.
2. Following case finding, a **data set** must be collected. The data set incorporates the type of information to be collected on each case, including any codes or classification systems to be used with the data collection. Usually, a data collection document or questionnaire is the final form for the data set. Surveillance for **epidemiologic** purposes involves an extensive data set (that is, relevant confounding exposures, age, sex, race, occupation) to permit a direct analysis from the data gathered through epidemiologic

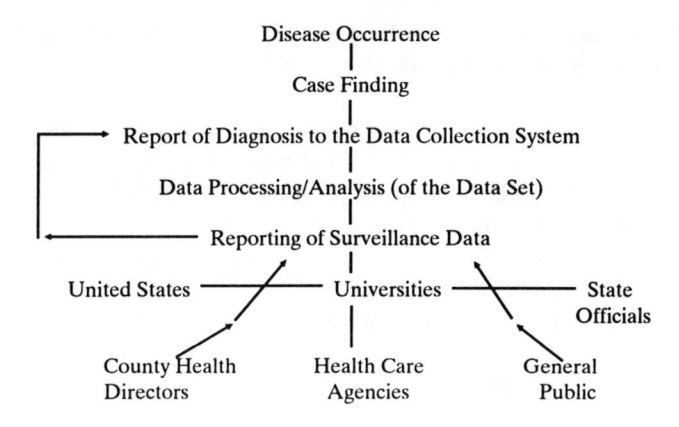

```
                    Disease Occurrence
                           |
                      Case Finding
                           |
         ┌──────→ Report of Diagnosis to the Data Collection System
         |                 |
         |    Data Processing/Analysis (of the Data Set)
         |                 |
         └──────────── Reporting of Surveillance Data
                  ╱        |        ╲
      United States ───── Universities ───── State
                  ╲        |        ╱       Officials
           County Health  Health Care   General
           Directors      Agencies      Public
```

FIGURE 5-1. Flow Chart of a Surveillance System and Reporting.

studies (for example, cohort, case-comparison, nested case-comparison, and descriptive). By contrast, **public health surveillance** involves a "minimal" data set primarily useful for monitoring disease occurrence; identified cases may then promote the development of epidemiologic studies (Thacker and Berkelman 1988; Houk and Thacker 1987).

Data items are usually those obtained from other medical documents, yet some may involve information obtained directly from the subject, either by interview or abstracting from a medical record. Data items may represent **primary variables** (for example, a direct measurement for the subject such as blood arsenic levels) or **secondary variables** such as information gathered on group characteristics (that is, county of residence, source of drinking water).

3. A **collection** system for case data must be developed. A **passive system** involving linkage of databases with information on persons with the disease of interest is used by many states. Although the **passive approach** is less expensive, it is subject to whatever ascertainment limitations were present in the original databases. Alternatively, the **active approach** is used by the Surveillance, Epidemiology, and End Results (SEER) registries of the National Cancer Institute. **Active** data collection in a surveillance program is similar to a cohort study. In the cohort study, initial data gathering ascertains exposure information; disease history accumulates as time passes. Surveillance activities are costly, as is the case with cohort studies. The **active** approach to the data collection process may involve two steps. First, an emphasis is placed on case finding. Then for special research purposes (as discussed under "Analytical Issues" in this chapter,

a more detailed data set may be collected for a subset of cases (that is, those suspected of being exposed or at risk).

In developing a surveillance program, emphasis should be placed on secular (general population trends, for example, the aging of the population here at the end of the twentieth century), or long-term (10–20 years) analyses of environmental exposures and disease occurrence.

As shown in Figure 5-2 and 5-3, you can easily see the changes occurring in mortality rates for selected cancers over time. Clearly, cancer of the lung, in females and males, shows an upswing in the 1970s and 1960s respectively. As shown in Figure 5-4, infant mortality in the United States has decreased for both black and white infants from 1950 to 1987. Black infant mortality peaked at about 48 deaths per 1,000 live births in 1950 and dropped to about 18 per 1,000 live births in 1987. White infant mortality peaked at about 28 deaths for every 1,000 live births in 1950, and fell to 8.4 deaths per 1,000 live births in 1987. Clearly, the rate of deaths for black infants is much higher than that of white infants. Congenital anomalies, sudden infant death syndrome, disorders relating to short gestation periods and low birth weight, and respiratory distress accounted for slightly more than 50 percent of all deaths of infants less than one year of age. The decrease in infant mortality was primarily due to the control of infectious diseases such as measles and whooping cough by immunization, and by the control of dysentery through improved environmental hygiene practices. The difference between infant mortality rates for white and black infants varied by cause, but for every risk factor, black infants were at increased risk (for example, for congenital anomalies, the leading cause of infant death, the difference between the rate for black infants (22.6/1,000 live births) and for whites (20.6/1,000 live births) was statistically significant). Health professionals feel that the increased infant mortality among blacks may be due to the lack of early prenatal care; however, infectious diseases are making a comeback now that compliance is less demanding concerning immunization levels.

Short-term biologic monitoring studies are an appropriate objective, but are probably several years away. Exposure measurements should include data on all relevant environmental media (that is, air, water, soil, food), as well as data on disease occurrence in selected exposed or **at-risk** groups and on the general population. Data collection should involve survey analysis (for example, case finding), and the collection of human tissue samples (for example, blood, urine, adipose) for monitoring pollutant residues in human tissue.

Surveillance data should provide guidance to managers of health care programs, support for regulatory program activities, and resources for scientists involved in health research. Modern surveillance systems use

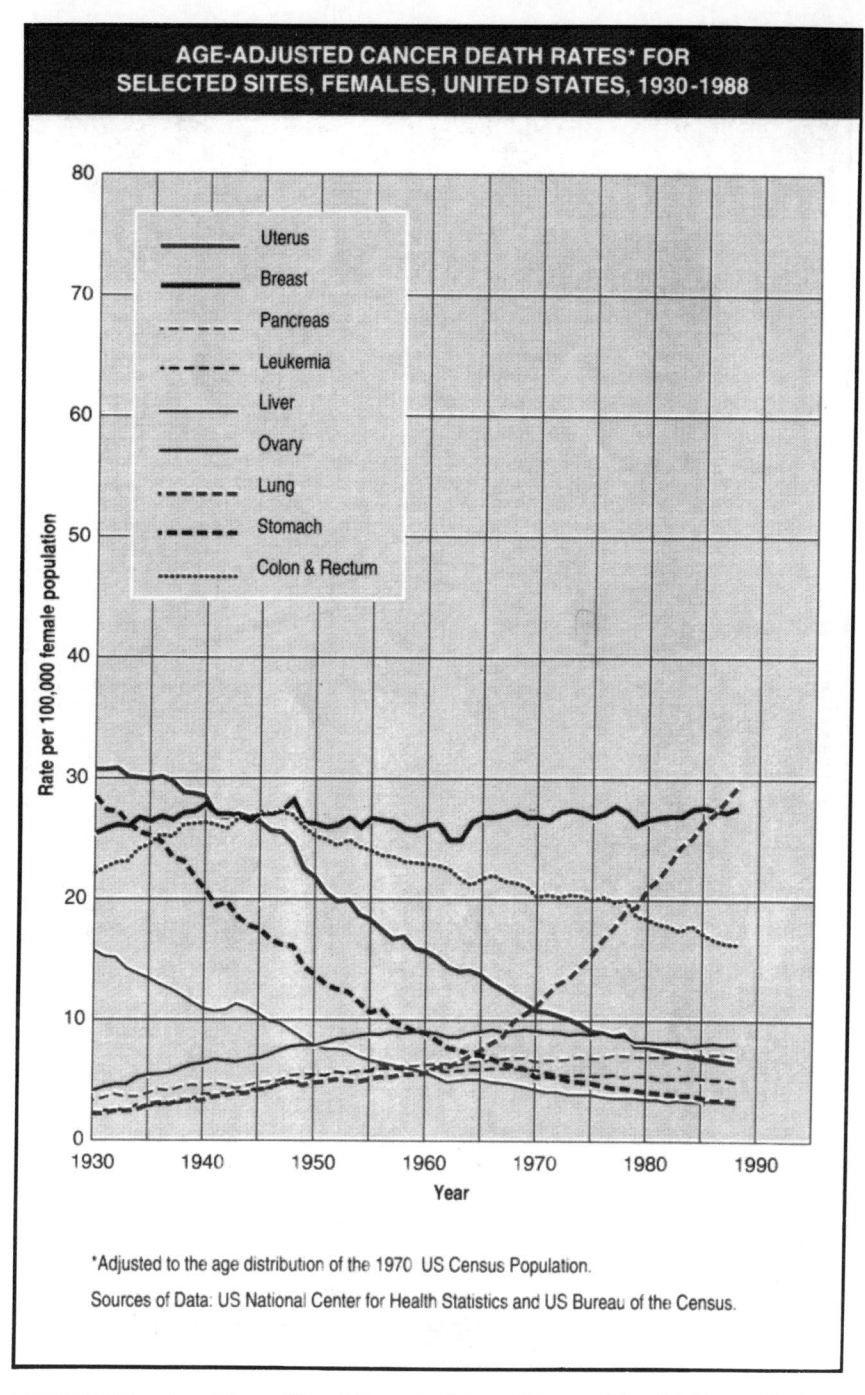

FIGURE 5-2. Age Adjusted Death Rates for Selected Cancers (Female). Source: Boring et al. 1992.

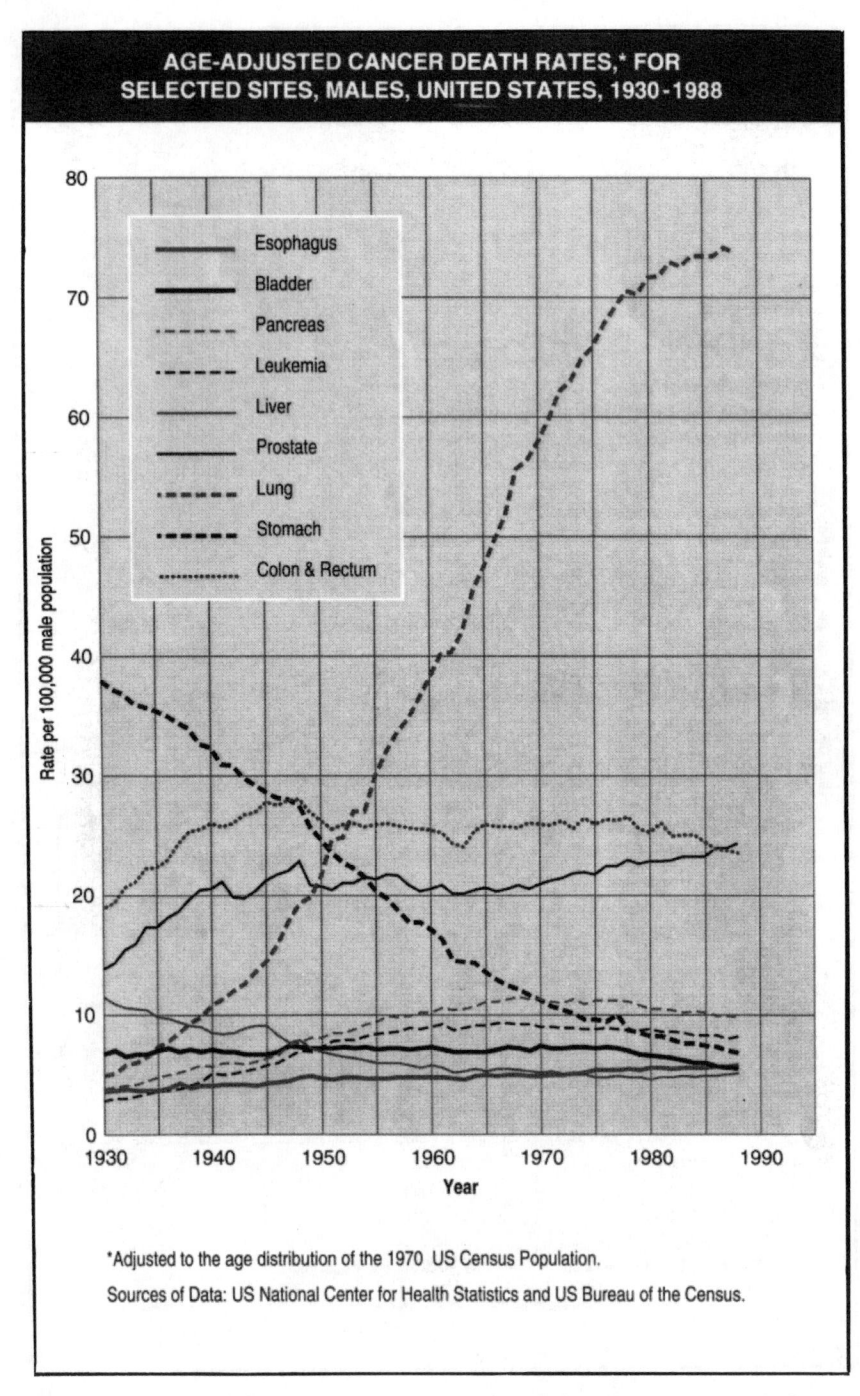

FIGURE 5-3. Age Adjusted Death Rates for Selected Cancers (Male). Source: Boring et al. 1992.

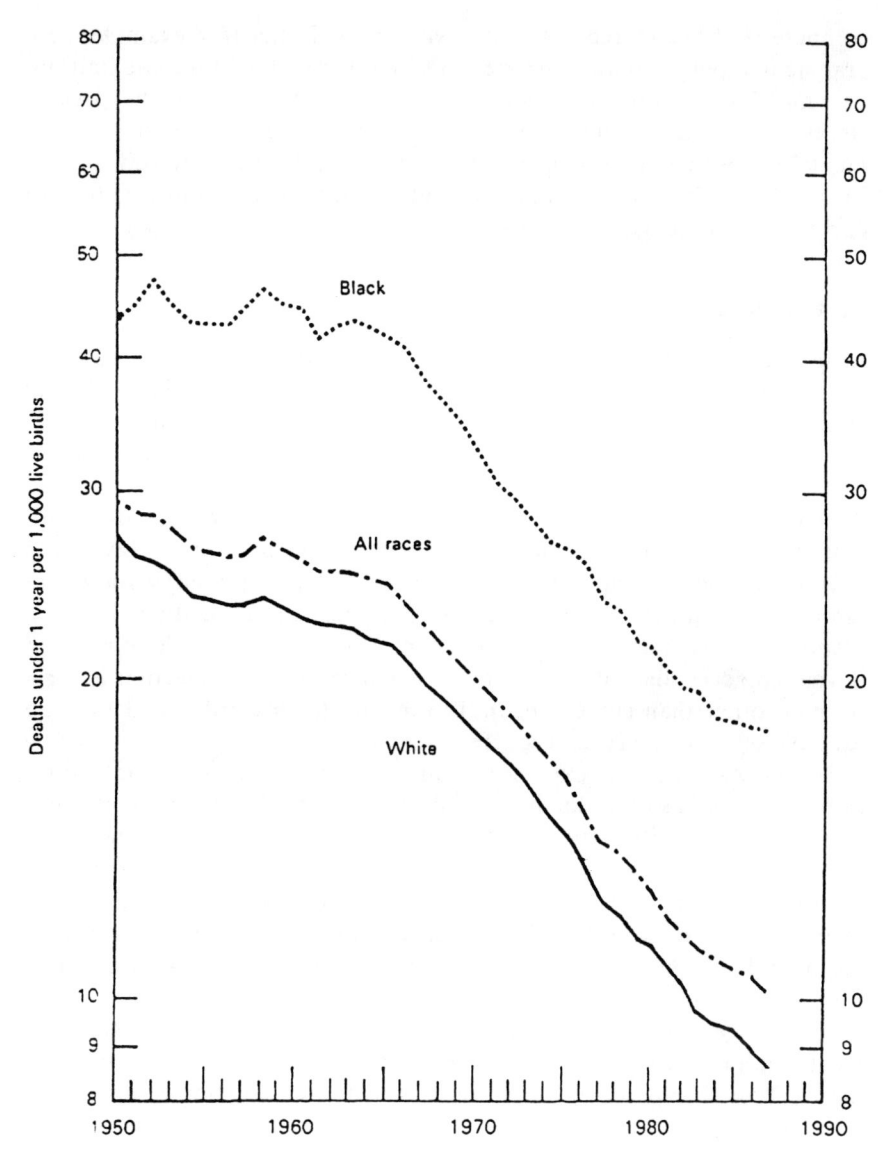

FIGURE 5-4. Infant Mortality Rates by Race: United States, 1950-1987. Source: Monthly Vital Statistics Report. NCHS. Vol. 38, No. 5. Supplement. Sept. 26, 1989.

computerized health records and environmental data (for example, geographic mapping, disease registries). Data analysis should include, but not be limited to, descriptive epidemiologic studies to evaluate the present **status** and developing **trends** of disease endpoints and associated environmental exposures (for example, data on pollutant levels in air, soil, water). This type of information will provide anticipatory research directed toward early problem recognition.

Data Sources

Data sources include hospital records; chronic and infectious disease registries; birth defects registries; hospital admissions; outpatient visits; well-child clinics; existing and developing environmental monitoring programs managed by local, state, and federal government agencies; and tissue data banks. Data from these sources are beneficial because they are collected and disseminated in a fairly standardized fashion. In population-based surveillance, data may be collected directly on exposed individuals, expressly for the purpose of study; this approach is known as developing **primary data.** A less direct approach may be used if cases of the disease to be studied can be identified through an existing data system. Such an approach is considered developing **secondary data** (i.e., the original reason for collection is for a purpose other than surveillance). The use of death certificate data is an example of a secondary data source.

In many states, laws expressly require reporting chronic or infectious diseases as well as reproductive effects data with public health significance (National Disease Registry 1990; Moldenbauer and Greve 1953). These laws compel institutions and health care professionals, primarily physicians, to report cases of selected diseases to a central agency or registry. To encourage compliance with the law, only a small amount of information is usually required. Forty-one states now have some level of population-based cancer surveillance and there are several birth defect registries as well (Cancer Registries 1990; Edmonds and Layde 1987). One can also use the National Death Index and local and state mortality registers to develop mortality information. Computerized data also aids with disseminating data to interested parties.

Data Linkage

State public health authorities have been encouraged to develop capabilities for conducting environmental health research by the Institute of Medicine Committee in its report on the Future of Public Health (Institute of Medicine 1988). Clearly, the linkage of chronic disease registry data with environmental databases (Schulte and Kaye 1988; Frisch et al. 1990) would facilitate such research. In addition, there are efforts underway by the Agency for Toxic Substances and Disease Registry (ATSDR) to develop a national disease registry (National Disease Registry 1990). This sort of nationwide database could lead to additional well-coordinated studies of populations exposed to hazardous materials, especially for exposures that cross state lines.

For example, data from the Iowa Cancer Center Registry was used to develop age-adjusted sex-specific cancer incidence rates 1969–1981 for towns with a population − 10,000 people and a public water supply from a single, stable ground source. These rates were related to levels of volatile organic compounds and metals found in the finished drinking water of these towns in the spring of 1979. Towns were dichotomized, < 0.5 µg/l and ≥ 0.5 µg/l. Results showed an association between nickel and cancers of the bladder (p = 0.005) and lung (p = 0.02) (Figure 5-5). The authors made a determination that the effects were noncausal because of the extremely low levels of nickel in the water, but suggested that they were indicators of possible anthropogenic contamination of other types of exposures.

Dissemination of Surveillance System Data

One critical characteristic of a surveillance system is the dissemination of data to appropriate authorities, including public health officials, physicians, health professionals, etcetera. Generally, data are disseminated through reports that vary from extensive to brief (for example, some systems will publish a concise annual report and make their existing database available to interested researchers). Periodically (for example, every 3–5 years), the system or registry will publish a detailed report with extensive analyses and narrative (for example, the Cancer Statistics Review).

ANALYTIC ISSUES

Of primary importance to the objective of environmental health surveillance is the ability to identify a change in disease status among selected populations. Surveillance systems provide data to appropriate authorities and health researchers for assessing disease occurrence in specific geographic areas, and for the identifying population subgroups. It is important to remember that there is an inherent difficulty in extrapolating disease incidence and mortality data from one locale to another because of differences in populations (for example, case finding and the identification of exposed populations is quite variable from one locale to another), and in environmental exposures.

Incidence and mortality rates also are quite variable (unstable) depending

92   Environmental Epidemiology

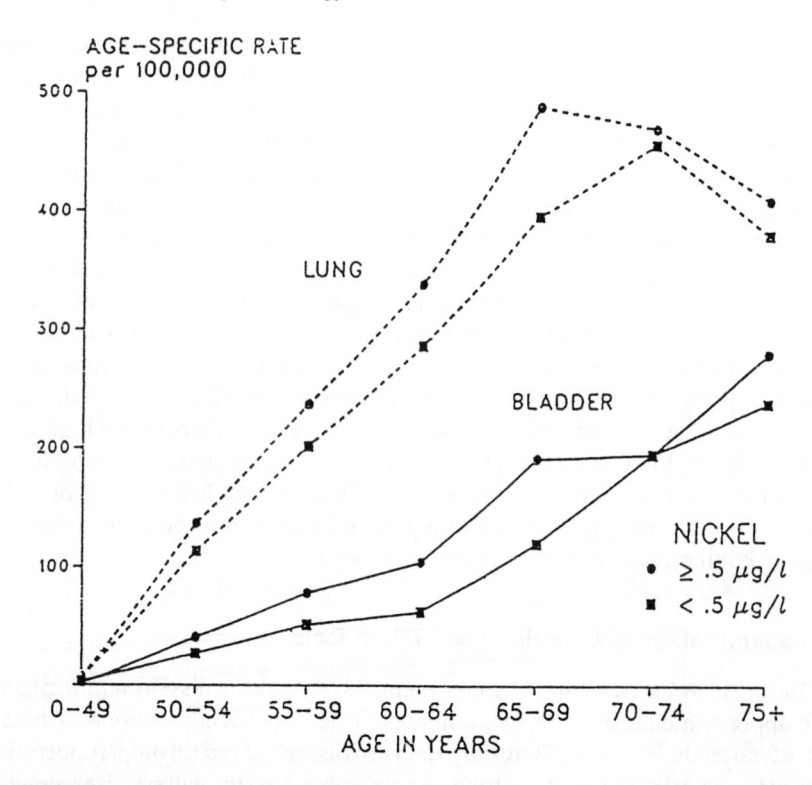

FIGURE 5-5.   Age-Specific Rates of Lung and Bladder Cancers Urinary Iowa Males by Level of Nickel in Finished Ground Water for the Years 1969-1981. Source: Isacson et al. Am. J Epidemiology 1985 Vol. 21, No. 6.

on the size of the denominator data available for estimation (that is, one or two cases of soft-tissue sarcoma in a population of only 100 people will certainly be more dramatic than one or two cases in a population of 1 million people). Other quantitative problems must also be considered, for example:

1. In Chapter 3, we mentioned the ecologic fallacy. This situation may occur where an increase in the rate of disease (for example, hypertension) is associated with an exposure to a particular group characteristic (for example, exposure to aircraft noise in a neighborhood under an airport landing pattern) when in fact the actual cause of disease might relate to individual characteristics (for example, age, sex, race, weight, genetic susceptibility).
2. Small expected values are a statistical, not an epidemiologic, issue. When one applies rates to a small population, not only are the rates unstable (for example, erratic), the numerators are often quite small, even fractionated

(for example, 0.1, or only one tenth a case is expected to occur). Therefore, when one finds just a few cases more than expected (for example, even one case can significantly skew the findings), there may be an observed risk measure that will appear to be quite dramatic (for example, tenfold occurrence for only one case). This circumstance is why Poisson confidence limits and statistical tests are most often used for rare event studies. Unlike normally distributed confidence limits or statistics, the Poisson distribution takes into account expected values that may be fractions. Poisson variables also have the desirable characteristic that their variation is comparable to the measure itself (that is, mean and variance are equal).

3. Even when conventional demographic characteristics (age, race, sex) are taken into account by rate adjustment, a severely skewed population may produce an artificially high rate. For example, in North Carolina the nonwhite population is primarily distributed among eastern countries. Ashe County is virtually all white (Table 5-1). Each year for the decade of the 1980s, Ashe County has experienced one case of a rare cancer. The case was always a white person, so that with appropriate age, race, sex adjustment, the rate was about that expected for the state as a whole. However, in 1986, the single case was an older, black resident in an age strata with only three other persons, resulting in an Age-, Race-, Sex-specific rate of 25,000 per 100,000 population. The corresponding Age, Sex, Race adjusted rate was 53.7/100,000 (almost a threefold increase over the statewide rate) that was completely skewed from the statewide rates. Upon learning of this dramatically elevated rate, community residents were apprehensive about the health risks posed by a local industry, and requested an epidemiologic study of this industry.

TABLE 5-1. Description of Ashe County Cases for Cancer X 1980–89.

Year	Case Description	Age-, Race-, Sex-Adjusted Rate	Standardized Mortality Ratio
1980	45 yo white male	13.5	0.66
1981	72 yo white female	19.7	0.97
1982	68 yo white male	20.6	1.01
1983	54 yo white male	23.5	1.16
1984	62 yo white female	15.4	0.76
1985	66 yo white male	18.9	0.93
1986	83 yo black female	53.7	2.64
1987	70 yo white female	16.1	0.79
1988	64 yo white male	15.8	0.78
1989	75 yo white female	19.3	0.95

Time Series Analysis

Identifying a **meaningful** increase in disease occurrence is both a sophisticated and a subtle problem. A well-established approach to conducting disease surveillance is to evaluate changes in disease rates over an extended period of time. As disease rates are monitored over time, a linear relationship may be developed and mathematically modeled. This mathematical model (termed a Fourier series) can be used to develop confidence limits on the expected, temporal disease pattern. Thus, within a single time period, if the observed disease rate exceeds the established **confidence limits,** an increase may be declared (Serfling 1963). This time-series approach has been the cornerstone of CDC's surveillance of influenza for decades. In recent years, the process of disease surveillance (for example, the estimation of disease rates among selected population groups) has become increasingly complex. This increased complexity is due to rapid population growth, which results in constantly shifting demographics: aging, migration of the population (from the snow belt to the sun belt), and industrial movement from one state to another. Consequently, in epidemiologic studies, the selection of an appropriate comparison population becomes more difficult, if not more critical. Furthermore, this process is complicated by the difficulty in determining whether observed increases in disease rates are a result of an environmental exposure (that is, are real), or just an unfortunate, random event. Often, the true meaning of the finding of an increased rate is determined by testing the null hypothesis that no increase exists. Such statistical testing issues lead logically to considerations for checking statistical significance. National Center for Health Statistics age-adjusted cancer mortality rates by primary cancer site, and 16-year trends for all races, males, and females, are shown in Table 5-2. Secular trends over the 16-year period suggest that cancers of the lung and bronchus, melanoma of the skin, multiple myeloma, and non-Hodgkins lymphoma have experienced the greatest increase among all cancer sites showing a positive increase during those years.

Spatial Analyses

Spatial analyses involve the evaluation of disease occurrence based on geographic patterns, or to monitor disease occurrence for evidence of geographic patterns. One issue with spatial analysis is whether geographic areas with similar rates (usually elevated) are geographically adjacent to one another.

TABLE 5-2. Age-Adjusted U.S. Cancer Mortality Rates and 16-Year Trends for Selected Sites, 1973–88.

Primary Cancer Site	Avg. Rate[λ] 73/74	Avg. Rate[λ] 87/88	Percentage Change	EAPC[§] 73/88
Lung and bronchus	35.4	48.2	36.1	2.2*
Melanoma of the skin	1.7	2.2	31.0	1.8*
Breast	14.9	15.3	2.4	0.3*
Brain and nervous system	3.7	4.1	11.0	0.5*
Non-Hodgkins lymph.	4.8	5.9	22.7	1.7*
Multiple myeloma	2.3	2.9	24.2	1.5*
Chronic lymphocytic leukemia	1.1	1.1	1.5	0.3*
All sites	162.6	171.4	5.4	0.4*

[λ]Rates are per 100,000 and are age-adjusted to the 1970 U.S. Standard Population.
[§]The estimated annual percent change over the 16-year interval.
* The EAPC is significantly different than zero ($p < 0.05$).
Source: Abstracted from Cancer Statistics Review 1973–1988. National Cancer Institute. NIH Pub. No. 91-2789, 1991.

Interpreting such an arrangement can be quite complex. New statistical software packages (for example, the Geographic Information System [GIS] technology; and the **CLUSTER** software programs) are available for use in centralized disease registries. Use of these, and other similar software will facilitate spatial investigations and studies of exposure from hazardous point sources. One of the more established methods for spatial analyses is called spatial auto-correlation (SAC) (Glick 1979). This technique performs a regression analysis for distances between geographic **cells** (usually counties) that have similar disease rates. The distance between similar cells is called **lags** (that is, the number of intervening cells with dissimilar rates). Spatial auto-correlation can be a very labor-intensive method, requiring considerable computer time. However, there are several advantages to SAC:

1. The degree of adjacency may be taken into account (that is, counties sharing a large, common border, vis-a-vis counties that barely touch).
2. Familiar statistical tests are used (z tables).
3. The cells may be weighted for demographic characteristics (for example, industrialization, urbanization, growth over time).
4. Disease rates may be adjusted for personal characteristics (for example, age, race, sex).

Another geographic approach, developed in England for the analysis of pediatric leukemia around nuclear power plants, is also available for use in geographic analyses (Beral 1990). The approach uses iterative processes, is computer intensive, and is suitable for studies with many subjects. The object of the approach is to study individual cases, and use concentric circles around cases to develop a **degree of intercept** statistic (see Chapter 4 for a more detailed discussion of spatial methods for the study of case-case proximity). The reason for mentioning this approach at this time is that it is a systematic approach intended to be applied in periodic analyses, expressly for use with cyclic studies of rare health events (that is, it is a surveillance system approach).

Because geographic areas are not usually symmetrical, one encounters many problems with studying counties (cells) with irregular borders. For example, **intracell** migration of cases is more a matter of convenient transportation to and from the place of residence than the amount of shared border space between counties. Intracell migration is often the case with suburban counties where large numbers of persons work in one county while living in a neighboring county. An innovative solution to the **unequal areas problem**, the Density Equalized Map Projection (DEMP) procedure, has been applied in California (Selvin et al., 1988). This approach adjusts the surface area of selected geographic areas for the underlying population distribution so that comparisons may be made more directly between areas.

For the *CLUSTER* software, two methods were chosen that are less computer intensive than those previously described, but that are also rather sophisticated tests for geographic aggregation (Aldrich 1990). Each of these methods processes the adjacency of cells with similar disease rates as the basis of analysis. One method was designed to use a hierarchy of solutions (that is, clusters of similar rates are sought using the data itself to guide the search) (Grimson et al. 1981).

A similar spatial technique was developed in Japan for studying cancer patterns in the provinces of that country (Ohno et al. 1979). The Japanese approach used the **nearest neighbor** solution, and it tests the observed pattern of adjacencies with a chi-square distribution. These methods are discussed under this heading because each was developed for use with a surveillance application (that is, repeated, continual tests over time).

In addition to studying a pattern of disease occurrence over time and in space, there is another approach to consider: the REMSA approach discussed in Chapter 4. REMSA is designed to answer the question, "What part of the available population **space** is being occupied by a subset of cases?" That is, REMSA searches for a small cell to test the possibility of a certain aggregation of cases occurring in that cell (Figure 5-6).

With this technique, individual population characteristics such as age, race,

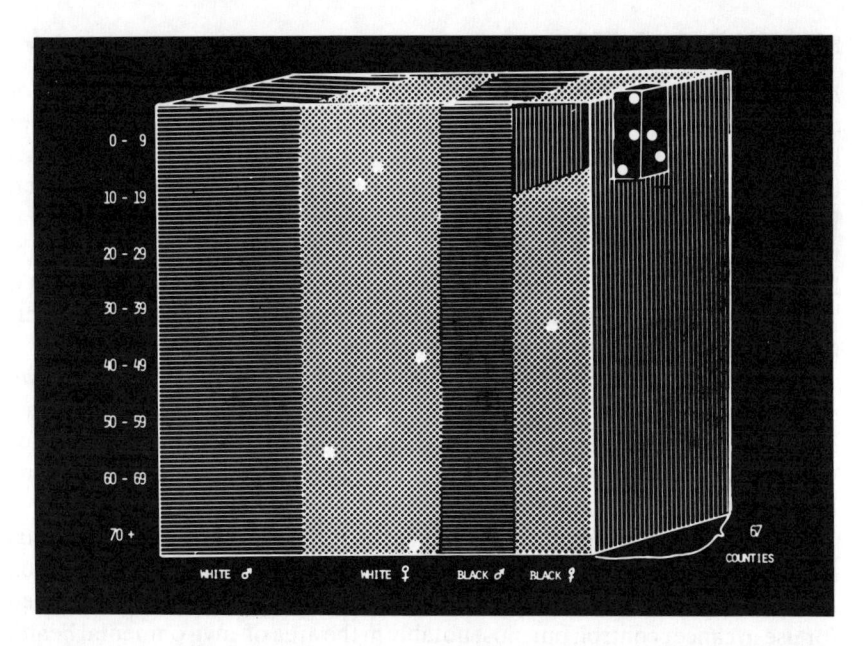

Surveillance Activities in Disease and Exposure Situations 97

FIGURE 5-6. REMSA CELL: A Multi-dimensional Cell for Evaluating the Probability of an Observed Case Aggregation.

and sex may be treated as independent probabilities. Consequently, a cluster of cases that occurs within a small population, a short time period, or a small geographic area can be detected, and all with the unifying characteristic that the aggregate is unlikely to be occurring by chance. This test uses a negative binomial solution to evaluate the probability of an unusual clustering of cases (see Figure 5-7). As with other methods mentioned in this section, the REMSA approach is meant to be used through time (continuously for surveillance).

$$\Pr = \frac{\binom{n}{k}}{m^n} (m - 1)^{n - k}$$

FIGURE 5-7. Negative Binomial—Formula With Explanation of Elements for Solution (Aldrich et al. 1985) m = the total number of cells available in the population space. k = the total number of cases of the event under study. n = the number of cases in the specific cell in question.

SENTINEL EVENTS

It is difficult for health professionals to explain to the public that one case of disease does not always equal another. For example, an increase of one case of lung cancer in a very large community with exposure to noxious air pollutants is of little statistical significance, especially if that case happens to be a smoker. However, one case of pediatric osteogenic sarcoma (a rare event) in a population living near a nuclear power plant can be a cause for substantive concern. Sherlock Holmes, in the *Case of the Redheaded League*, told Dr. Watson that "The more bizarre a thing is, the less mysterious it is likely to prove." This adage is a prelude to what is called sentinel event reasoning.

The biologic meaning of certain events may be construed to be greater than other events because of recognized characteristics that are known or suspected about the disease etiology. **Sentinel events** can provide an early warning for potential health problems, thereby permitting health authorities to intervene before they become a real problem.

The use of sentinel events in surveillance is borrowed from occupational medicine where they are used to identify circumstances where intervention is warranted (Rutstein et al. 1984). The use of sentinel events can be applied for use in cancer control, but most notably in the area of environmental health (Rothwell et al. 1991). The required elements in a surveillance program for the use of sentinel events are:

1. A listing of events,
2. A case reporting system, and
3. Statistical methods for declaring an increase in the rate of a disease based on one or very few occurrences.

The sentinel event approach has been used successfully in evaluating a national study of the geographic distribution of deaths related to environmental causes (Wagener and Buffler 1989). Large industries are developing disease and medical surveillance systems based on the use of sentinel events, and the National Institute of Occupational Safety and Health is presently operating a sentinel event based surveillance system (Joiner 1982; Ehrenberg 1987).

What should be considered as a sentinel event? Returning to the Holmes quote, Miller (1981) notes that **virtually every known carcinogen and teratogen has first been recognized by an alert clinician.** The clinician in each case was knowledgeable about disease occurrence, so that when a sentinel event occurred, they noticed it. Thus, rare diseases and diseases occurring in unusual population groups would be at the forefront of a list (Table 5-3) of sentinel health events (Garfinkel 1987), and conditions (Table 5-4); "criteria" are the usual groups for listed events. Furthermore, sentinel events should

TABLE 5-3. Sample Listing of Sentinel Health Events (Criteria and ICD-9-CM Codes).

Liver carcinoma (nonsmoker/nondrinker)	155.0
Glioblastoma	191.0
Amelanotic melanoma	172.0
Pediatric solid tumors (especially germ cell origin)	any site
Genitourinary cancers in children	179.0–189.0
Exotic lymphoma cell types	196.0
Exotic leukemias (or uncharacteristic age groups)	200.0–208.0
Bladder, lung, or upper respiratory cancer (nonsmoker)	188.0, 162.0, 146.0–149.0
Midline or septal birth defects	749.0, 745.0

Source: Aldrich, et al., 1989.

TABLE 5-4. Sample Listing of Sentinel Health Conditions (Criteria and ICD-9-CM Codes).

Unusual allergies	
Unusual neurological symptoms	
Idiopathic hematuria	599.7
Persistent, unresolved rashes	692.9
Persistent, idiopathic nasopharyngitis	472.2
Extreme liver functions in a nonsmoker	
Extreme renal functions in a nonsmoker	

Source: Aldrich, et al., 1989.

have temporal validity (that is, most cancers are not useful sentinel events because their latency is too long, in the range of 15–30 years).

Sentinel events require practitioners to know enough about suspect exposures that they are able to anticipate sequelae (that is, the **index** case of one of these diseases would signal the need for remedial action). To develop a system based on sentinel events, it is absolutely necessary that a mechanism be available to provide **early warning of an unusual health event,** or the identification of an unusual pattern of events. It is also necessary to have a method for detecting unusual aggregates of disease so that statistically significant disease clusters can be identified. Although these objectives are consistent with the operation of a surveillance system, currently the approach to the use of sentinel events is being driven by an inefficient process of disease cluster reports (Enterline 1985; Schulte et al. 1987; Rothman 1990).

Monitoring of adverse reproductive outcomes, neurologic conditions and acute diagnoses is also needed (Miller 1981; Buffler and Aase 1982). As our understanding of genetic factors associated with susceptibility to hazardous exposures grows, there is the potential for monitoring specific biologic markers and identifying individuals with genetic predisposition to disease (Nicholson 1984; Hulka et al. 1990).

MODIFICATIONS OF THE SURVEILLANCE RESEARCH DESIGN

There are three principal modifications of the approach:

1. Classification by active exposure status such as active monitoring (for example, film badges with ionizing radiation), and a periodic assessment of health status (for example, annual physical examinations) (Newcomb et al. 1990). This form of surveillance is termed **medical surveillance.** Follow-up is maintained on all persons in the cohort (for example, who begin the study), even if their exposure ceases. Fewer subjects are followed, but they are studied longer and more closely, thus raising the cost per person.
2. Retrospective cohort studies conducted with community registries are more passive because the exposed and nonexposed persons are not actually defined as such during the follow-up period. This approach is sometimes identified as **public health surveillance.** A **stable** study population is assumed so that an individual who leaves the community is thought of as being replaced by another individual moving into the community. The entire community, or all exposed persons, are maintained under surveillance and data is collected regarding exposure status only when the selected diagnosis occurs. This technique is a **nested** design and the study could last from one to several years, depending on the rarity of the disease. Because many more people may be followed, the cost per subject is lower. Many states are using this approach to study communities around hazardous waste sites and even large industries (ATSDR 1988; Bond et al. 1988).
3. The Agency for Toxic Substances and Disease Registry has several exposure registries (ATSDR, 1988) in place. Persons are followed in these registries as in the previous modification except that assessment of more acute symptoms and health effects is possible. At this time, participants are asked to complete a health inventory annually. The active data collection and targeted data set lead this approach to be termed **epidemiologic surveillance.** Again, there are specialized analytic methods expressly designed for studying events that occur over time, and statistical consultation is prudent.

Because surveillance analyses can be so complex, it is a sound strategy to develop models of the expected disease rates and to prepare thresholds of observed cases/rates that will represent statistical significance (see Chapter

3). In surveillance activities, you must always be alert for a purely random increase in disease rates **(false positive findings)**. A statistical solution can be built into the decision threshold to permit one **false positive** per a series of tests (Hardy et al., 1990; Chen 1978). Also, a sequential approach may be used so that an increase will first trigger an **ALERT** that a threshold has been passed, then if the elevated occurrence continues for a second interval, intervention action can be taken (Glasser 1986). This same approach permits a single, very high increase to signal **ACTION** based on one time period alone (Hardy et al. 1990) (Table 5-5). A tiered response mechanism like this (**ALERT** then **ACTION**) is a useful enhancement for surveillance system analyses.

SUMMARY

Surveillance systems provide valuable information on the magnitude and etiologic factors associated with infectious and chronic diseases. These pro-

Table 5-5. Summary Table of the Number of Deaths Required (Or the Magnitude of SMR Required) for an Alert* or Action to be Taken for Various Values of the Expected Number of Deaths.

Expected Number Of Deaths	Number of Deaths For an Alert	For an Action
0.050	1 (20)**	3 (60)
0.100	2 (20)	3 (30)
0.200	2 (10)	4 (20)
0.400	2 (5)	4 (10)
0.500	2 (4)	5 (10)
1	3 (3)	6 (6)
2	5 (2.5)	9 (4, 5)
4	8 (2)	12 (3)
5	9 (1.8)	14 (2.8)
10	15 (1.5)	22 (2.2)
15	21 (1.4)	29 (1.93)
20	26 (1.3)	36 (1.8)
25	33 (1.32)	43 (1.72)
30	38 (1.27)	49 (1.63)

* Action and alert levels correspond to $p_2 = 0.001$ and $p_1 + p_2 = 0.09$ for a two-year error level of 0.01.
** ( ) Corresponding SMR associated with the specified expected number of deaths and observed number of deaths.
Source: Hardy, R. J. 1983. *Monitoring for Health Effects of Low-Level Radioactive Waste Disposal: A Feasibility Study.*

grams provide a database for early investigation of infectious disease outbreaks and for the investigation of etiologic factors associated with chronic disease. Surveillance information also provides support to programs designed to remedy conditions that may contribute to the onset and transmission of infectious diseases (for example, increased immunization), and the diagnoses and prevention of chronic disease (for example, mammography for female breast cancer detection).

In developing disease surveillance systems we need research methods based on prospectively identifying meaningful rare event aggregates (Miller 1981; Aldrich et al. 1983). These methods would be most productive if they were designed to operate with sentinel events (Aldrich et al. 1989). However, disease surveillance systems are expensive, and although they can effectively contribute human information to the risk assessment process (see Chapter 10), thereby reducing uncertainties associated with **risk assessment** (Beral 1985; Schulte et al. 1987), the strategic and political wisdom to financially support such systems is often lacking (Brownlea 1981; Gaugh 1987).

ASSIGNMENTS

1. Look at the Ashe County Data in Table 5-1 and use the data in Table 5-5 to determine if an increased rate of Cancer X occurred in 1986. Write your answer as a short narrative.
2. How would you relate the concept of an ALERT and ACTION levels to a supervisor who had just joined your department and was unfamiliar with statistics (You may want to read Hardy et al. 1990).
3. Compare population surveillance for disease cases versus surveillance of exposed individuals. Are they different? If so, how do they differ? What is the impact on the total surveillance system approach by unit of study?

REFERENCES
Aldrich, T. E. 1990. *CLUSTER: User's Manual for Software to Assist with the Investigation of Rare Health Events*. Published by the Agency for Toxic Substances and Disease Registry, Atlanta, Georgia.
Aldrich, T. E., Meyer, R. E., Qualters, J., and Atkinson, D. A. 1989. "Rare Health Events as Sentinels of Environmental Contamination." In *Proc. Public Health Conference on Records and Statistics*, DDHS Pub. No. 90-1214:323–6.
Aldrich, T. E., Wilson, C. C., and Easterly, C. E. 1986. "Population-Surveillance for Rare Health Events." Proceedings of the 1985 *Public Health Conference on Records and Statistics,* DHHS Pub. No. (PHS) 86-1214:215–220.
Aldrich, T. E., Garcia, N., Zeichner, S., Berger, S. 1983. "Cancer Clusters: A Myth or A Method." *Medical Hypotheses* 12:41–52.
ATSDR. 1988. Agency for Toxic Substances and Disease Registry. *Policies and*

Procedures for Establishing a National Registry of Persons Exposed to Hazardous Substances. National Exposure Registry, Atlanta, Georgia.

Beral, V. 1990. "Childhood Leukemia Near Nuclear Power Plants in the United Kingdom: The Evolution of a Systematic Approach to Studying Rare Disease in Small Geographic Areas." *American Journal of Epidemiology* (132-Supplement):S63–S68.

Boring, C. C., Squires, T. S., Tong, T. 1992. "Cancer Statistics." *CA: Cancer Journal for Clinicians* 42:28–29.

Brownlea, A. 1981. "From Public Health to Political Epidemiology." *Social Science and Medicine* 15D:57–67.

Buffler, P. A., and Aase, J. M. 1982. "Genetic Risk and Environmental Surveillance: Epidemiologic Aspects of Monitoring Industrial Populations for Environmental Mutagens." *Journal of Occupational Medicine* 24(4):305–14.

Bond, G. G., Austin, D. F., Gondek, M. R., Chiang, M., and Cook, R. R. 1988. "Use of a Population-Based Tumor Registry to Estimate Cancer Incidence Among a Cohort of Chemical Workers." *Journal of Occupational Medicine* 30(5):443–48.

Cancer Registries 1990. "Utility of Cancer Registries Varies." *Public Health Macroview* 3(5):2–3.

Chen, R. 1978. "A Surveillance System for Congenital Malformations." *Journal of American Statistical Association* 73:323–27.

Edmonds, L. D., and Layde, P. M. 1987. "Congenital Malformation Surveillance: Two American Systems." *International Journal of Epidemiology* 10:247–51.

Ehrenberg, R. 1987. "Sentinel Events Notification System for Occupational Risks (SENSOR)." In *Proceedings of the Workshop on Needs and Resources for Occupational Mortality Data.* Centers for Disease Control. DHHS Pub. No. (PHS) 88-1463:448–49.

Enterline, P. E. 1985. "Evaluating Cancer Clusters." *American Industrial Hygiene Association Journal* 46:10–13.

Frisch, J. D., Shaw, G. M., and Harris, J. A. 1990. "Epidemiologic Research Using Existing Databases of Environmental Measures." *Archives of Environmental Health* 45(5):303–7.

Garfinkel, L. 1987. "Cancer Clusters." *CA: A Cancer Journal of Clinicians* 37(1): 20–25.

Gaugh, M. 1987. "Environmental Epidemiology: Separating Politics and Science." *Issues in Science and Technology* Summer:20–31.

Glasser, J. H. 1986. "Health Statistics Surveillance Systems for Hazardous Waste Disposal." *Proceedings from the 1985 Public Health Conference on Records and Statistics.* DHD Pub. No. (PHS) 86-1214:221–24.

Glick, B. 1979. "The Spatial Autocorrelation of Cancer Mortality." *Social Science and Medicine* 13D:123–30.

Grimson, R. C., Wang, K. C., and Johnson, P. W. C. 1981. "Searching for Hierarchial Clusters of Disease: Spatial Patterns of Sudden Infant Death Syndrome." *Social Science and Medicine* 15:287–93.

Hardy, R. J., Schroeder, G. D., Cooper, S. P., Buffler, P. A., Prichard, H. M., and Crane, M. 1990. "A Surveillance System for Assessing Health Effects From Hazardous Exposures." *American Journal of Epidemiology* 132:S32–S42.

Hattis, D. B. 1986. "The Promise of Molecular Epidemiology for Qualtitative Risk Assessment." *Risk Analysis* 6(2):181–93.

Houk, V. N., and Thacker, S. B. 1987. "Registries: One Way to Assess Environmental Hazards." *Health and Environmental Digest* 1:5–6.

Hulka, B. S, Wilcosky, T. C., and Griffith, J. D. 1990. *Biological Markers in Epidemiology*. New York: Oxford University Press.

Institute of Medicine. 1988. *The Future of Public Health*. Pub. National Academy of Science, New York.

Isacson, P., Bean, J. A., Splinter, R., Olson, D. B., and Kohler, J. 1985. "Drinking Water and Cancer Incidence in Iowa. III. Association of Cancer With Indices of Contamination." *American Journal of Epidemiology* 121(6):863.

Joiner, R. L. 1982. "Occupational Health and Environmental Information Systems: Basic Considerations." *Journal of Occupational Medicine* 24(10):863–66.

Ohno Y Aoki K and Aoki N. 1979. "A Test of Significance for Geographic Clusters of Disease." *International Journal of Epidemiology* 8:273–81.

Mausner, J. S., and Bahn, A. K. 1974 *Epidemiology: An Introductory Text*. Philadelphia, PA: WB Saunders Company.

Miller R. W. 1981. "Area Wide Chemical Contamination: Lessons from Case Histories." *Journal of the American Medical Association*. 245:1548–51.

Moldenbauer, R. M., and Greve, C. H. 1953. "General Regulatory Powers and Duties of State and Local Health Authorities." *Public Health Reports* 68:434–438.

National Disease Registry. 1990. *Final Report of the Panel on the National Disease Registry*. Atlanta, GA: Published by the Agency for Toxic Substances and Disease Registry.

Newcomb, P. A., Love, R. R., Phillips, J.L., and Buckmaster, B.J. 1990. "Using a Population-Based Cancer Registry for Recruitment in a Pilot Cancer Control Study." *Preventive Medicine* 19:61–65.

Nicholson, W. J. 1984. "Research Issues in Occupational and Environmental Cancer." *Archives of Environmental Health* 39(3):190–202.

Rothman, K. J. 1990. "A Sobering Start for the Cluster Busters' Conference." *American Journal of Epidemiology* 132-Supplement S6–S13.

Rutstein, D. D, Mullan, R. J., Frazier, T. M., Halperin, W. E., Melius, J. M., and Sestito. 1984. "Sentinel Health Events (Occupational): A Basis for Physician Recognition and Public Health Surveillance." *American Journal of Public Health* 39(3):1054–62.

Schulte, P. A., Ehernburg, R. L., and Singal, M. 1987. "Investigation of Occupational Cancer Clusters." *American Journal of Public Health* 77(1):52–56.

Schulte, P. A. and Kaye, W. E. 1988. "Exposure Registries." *Archives of Environmental Health* 43(2):155–61.

Selvin, S., Merril, D., Schulman, J., et. al. (1988) Transformations of maps to investigate clusters of disease. *Social Science in Medicine*. 26:215–21.

Serfling, R. E. 1963. "Methods for Current Statistical Analysis of Excess Pneumonia-Influenza Deaths." *Public Health Reports* 78:494–506.

Thacker, S. B. and Berkelman, R. L. 1988. "Public Health Surveillance in the United States." *Epidemiologic Reviews* (10):164–90.

Wagener, D. K., and Buffler, P. A. 1989. "Geographic Distribution of Deaths Due to Sentinel Health Events (Occupational) Causes." *American Journal of Industrial Medicine* 16:355–72.

6

Characterizing Human Exposure

Jack Griffith and Tim E. Aldrich

OBJECTIVES

This chapter will:

1. Discuss the magnitude of chemical contamination in the ambient environment.
2. Outline the steps involved in characterizing human exposure.
3. Review the types of human exposure and methods of measurement.
4. Review the issues involved in assessing health risks associated with environmental exposure to toxic chemicals in air, water, and soil.

CHARACTERIZING HUMAN EXPOSURE

Characterizing human exposure is the process of measuring or estimating the intensity, frequency, and duration of human exposure to environmental agents from source to organ site. The process of characterizing exposure involves several facets, including identifying the source, monitoring ambient levels in all environmental media (that is, air, water, soil) and understanding the transport mechanism; the impact of environmental bioaccumulation on potential dose; and finally, the ability to quantify dose at the tissue or organ site level. Clearly, if epidemiologists had a choice, they would prefer to have exposure quantified at the organ site, but this is seldom possible.

Steps Involved in Characterizing Exposure

Characterizing exposure describes the magnitude, duration, and route of exposure on populations by age, sex, race, and size and involves the following steps:

1. Identifying and measuring the pollutant,
2. Quantifying the magnitude of the pollutant in the environment,
3. Identifying the source of the pollutant,
4. Identifying the exposure media,
5. Identifying the pollutant's means of transport,
6. Identifying the pollutant's chemical and physical properties and means of transformation,
7. Identifying the pollutant's routes of entry into the body,
8. Identifying the pollutant's intensity and frequency of contact,
9. Identifying the pollutant's spatial and temporal concentration patterns, and
10. Estimating the total exposure to different compounds and mixtures (NRC 1991).

Human exposure characterization has proven difficult because human data are often not available. This is because people are exposed to complex mixtures of chemicals, and because environmental monitoring methodology is difficult and expensive. Although records of chemical use can serve as a surrogate of exposure, such information is seldom available. Today, efforts are underway to estimate exposure with mathematical and pharmacokinetics modeling.

In the **classic** epidemiologic approach (Chapter 3), human exposure is estimated from existing records such as employment histories, medical records, or questionnaire data recalled by the study participant. Unfortunately, relying solely on questionnaire data or existing records will likely lead to some degree of **misclassification** of the study participants, and may even result in selection bias in the study. Exposure characterization designed to measure concentrations of exposure in the ambient environment, and when possible, the concentration of chemical levels in human tissue and fluids, is truly the **gold standard** in environmental epidemiology.

Human Exposure

Human exposure to toxic chemicals occurs when persons mix, load, and apply chemicals; when they enter areas where chemicals are manufactured or used; or when there are excess levels of toxicants in the ambient environment (that is, air, soil, water). When attempting to characterize human exposure, always be aware of the potential for spatial and temporal variation of the pollutant (for example, lead levels in ambient air will likely be higher near a heavily traveled throughway, and would be expected to peak during rush hour). One should also understand that humans are generally very peripatetic, that is, we move from place to place, making it difficult to relate site-specific measurements to individual exposures. Interestingly, people tend to stay inside more than outside the home or workplace. In fact, it has been estimated that people spend as much as 70 percent of their time indoors (Ferris et al., 1988). This is important in that indoor levels of recognized ambient pollutants can far exceed outdoor levels. Sexton et al. (1983) measured the composite 24-hour annual indoor concentration for gas cooking homes in Portage, Wisconsin. They found that the mean indoor concentration for nitrogen dioxide (NO_2), was 65 $\mu g/m^3$ while the mean outdoor concentration was 15 $\mu g/m^3$. In fact, the values for indoor NO_2 exposures in 3 percent of the homes exceeded 100 $\mu g/M^3$, the National Ambient Air Quality Standard (NAAQS). The investigators determined that when the indoor and outdoor exposure values were combined, 67 percent of the variation in week-long integrated (the dose or fraction of dose retained from a previous time period will be added to the present time period) personal NO_2 exposures for adults and children could be explained.

Clearly, in characterizing human exposure for epidemiologic studies, indoor exposures (home and workplace) must be taken into consideration. One must also consider the value of the personal monitor as opposed to the use of a site monitor. For example, an air monitoring device that will be described on the following pages provides excellent data on the amount and kinds of pollutants in the ambient air in the work or home environments. However, any assessment of exposure gathered from this site monitor must be considered to be group exposure, rather than individual exposure. Thus, in attempting to determine a relationship between exposure and health, analyses must focus on the group—not on the individual. On the other hand, where personal monitors are available, it will be possible to directly measure individual exposure and reduce the potential for misclassification in the establishment of a study cohort.

Since environmental epidemiology often involves the association of complex mixtures of chemicals to relatively rare disease endpoints, frequently with long induction and latency periods, it is absolutely necessary that exposure be measured with as much validity (sensitivity and specificity) as possible. For the purpose of characterizing exposure, sensitivity is defined as the ability to correctly identify persons with the exposure, while specificity is defined as the ability to currently identify those persons who are not exposed (Griffith et al., 1989).

MAGNITUDE OF THE POTENTIAL PROBLEM

There are more than 65,000 chemicals presently in use in the United States, and another 1,000 introduced yearly (about two dozen of these chemicals are known human carcinogens (Goldstein 1988; Vainio et al., 1981). Agricultural workers and others are exposed to a variety of chemicals, including fertilizers, solvents, paints, fuels, and pesticides (over 1,400 active ingredients are formulated into more than 45,000 pesticide products). Because more than 2.29 billion pounds (approximately 1 billion kg) of active ingredient (component in the mixture that activates the toxic reaction) pesticides are used in the United States each year (Laughlin and Gold 1988), pesticide exposure is one of the most common environmental exposures. Over the long term, more than 200 million pounds of active ingredient pesticides were applied from 1971 to 1981.

We also manage to produce more than 6 billion tons of toxic wastes annually (almost 50,000 lbs per person) (OTA 1989). In 1988, the U.S. EPA estimated that the amount of toxic waste managed by more than 3,000 facilities was about 275 million metric tons (EPA 1988). The National Priority List (NPL) of Superfund sites lists more than 400,000 hazardous waste sites in the United States (NRC 1991). Almost four million people live within one mile of a toxic waste site, while almost 40 million live within four miles (Figure 6-1).

The EPA evaluated the composition of leachates from 13 toxic waste sites located throughout the United States. Only 4 percent of the total organic carbon (TOC) in the leachate was characterized by gas chromatography and mass spectroscopy according to chemical structure. More than 200 separate compounds were identified in the 4 percent fraction (for example, 42 organic acids, 43 oxygenated and heteroaromatic hydrocarbons, 39 halogenated hydrocarbons, 26 organic bases, 32 aromatic hydrocarbons, 8 alkanes, and 13 metals), leaving 96 percent of the organic carbon unknown. Therefore, although we see significant potential toxicity in the chemical compounds identified, the 4 percent only represents a small fraction of the overall organic contribution to the waste site.

As shown in Figure 6-2, the physical and biological routes of exposure are numerous and complex. All the environmental media (water, soil, and air) are potential reservoirs for chemical contamination. Humans have direct contact with toxicants by eating contaminated food, drinking contaminated water, and breathing contaminated air. In this chapter, we will discuss the various sources of contamination, chemical transport from the environment to the human, means of exposure that humans encounter, and finally absorption of toxic chemicals into human tissue.

SOURCES OF ENVIRONMENTAL CONTAMINATION

Groundwater Contamination

Millions of tons of hazardous wastes are currently leaching into groundwater in areas that pose a potential risk to people who rely on groundwater (for ex-

Characterizing Human Exposure 109

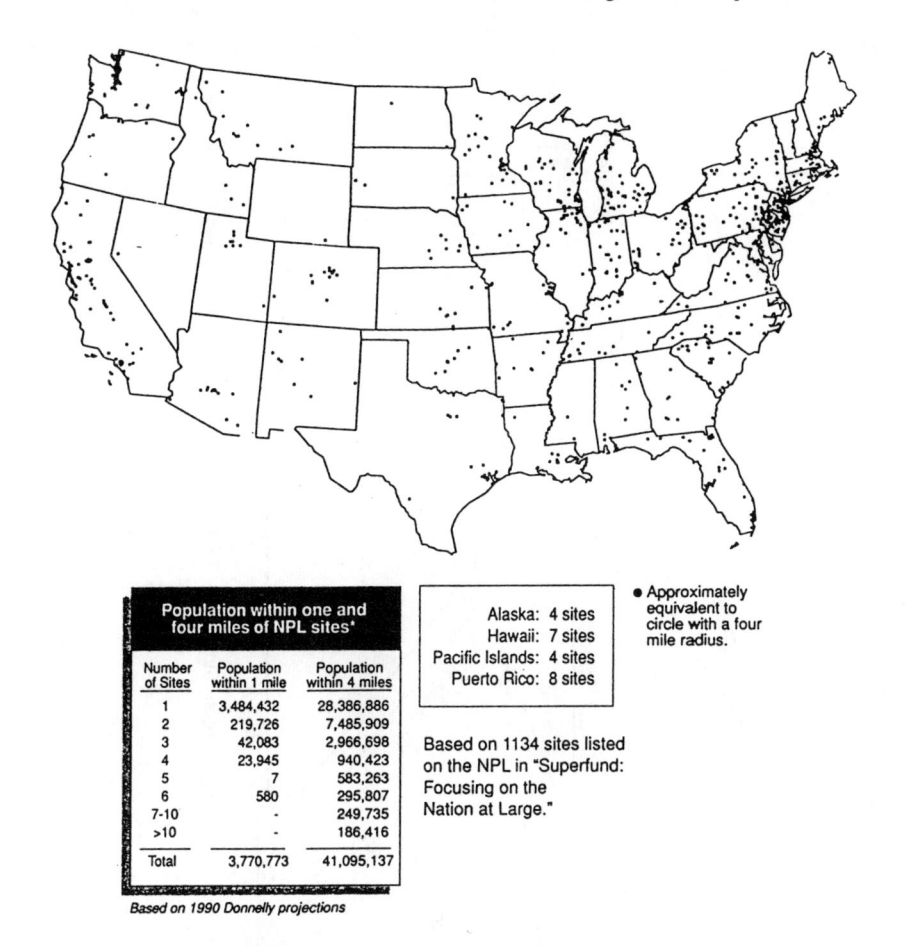

Population within one and four miles of NPL sites*		
Number of Sites	Population within 1 mile	Population within 4 miles
1	3,484,432	28,386,886
2	219,726	7,485,909
3	42,083	2,966,698
4	23,945	940,423
5	7	583,263
6	580	295,807
7-10	-	249,735
>10	-	186,416
Total	3,770,773	41,095,137

Alaska: 4 sites
Hawaii: 7 sites
Pacific Islands: 4 sites
Puerto Rico: 8 sites

• Approximately equivalent to circle with a four mile radius.

Based on 1134 sites listed on the NPL in "Superfund: Focusing on the Nation at Large."

Based on 1990 Donnelly projections

FIGURE 6-1. National Priority List Sites and Population Resident Within One and Four Miles. Source: Environmental Protection Agency, Office of Solid Waste and Emergency Response, 1991.

ample, 95 percent of all rural people, and 50 percent of U.S. population in general) as their main source of drinking water. Clearly, chemical contamination of drinking water is a concern in areas of the country where chemical leachate has been known to pollute drinking water and has been associated with adverse health effects. Sources of drinking water may be contaminated in many ways (for example, chemical residues in the ambient air that find their way to streams, rivers, and lakes; residues from chemical waste sites that flow directly to the water source [reservoir or river]; or leachate from chemical storage, use, or spills that finds its way into groundwater). A major source of drinking water contamination is through leachate from hazardous wastes disposal sites.

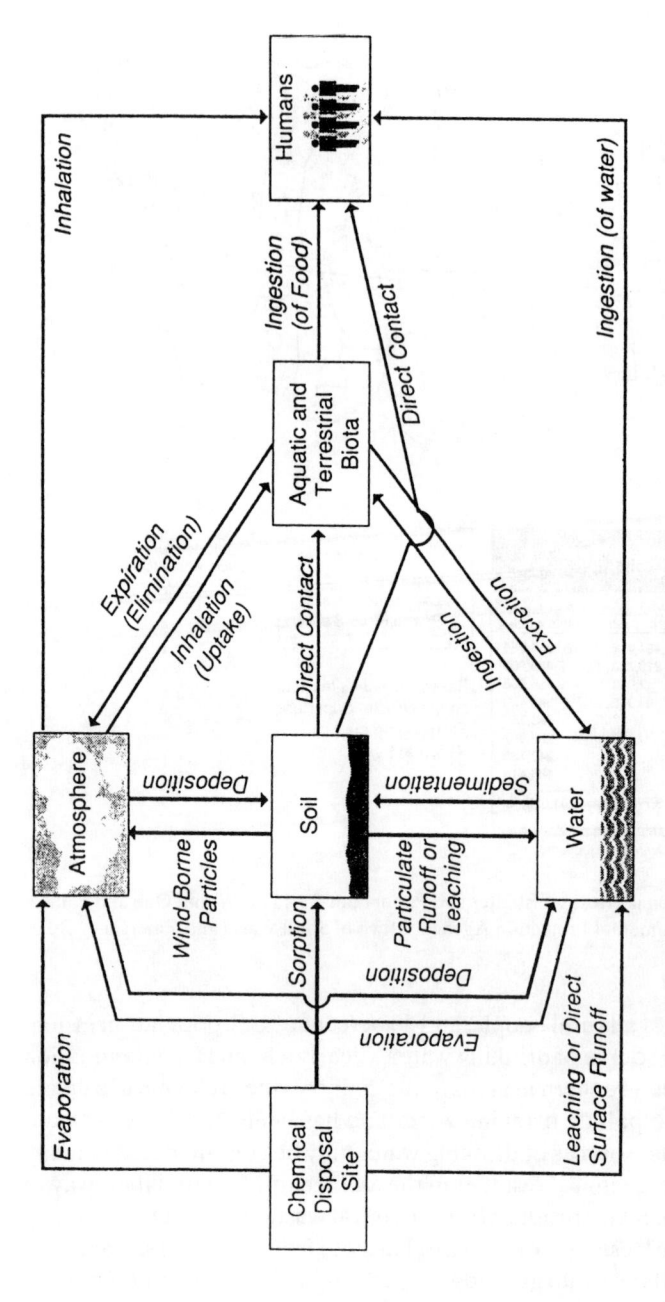

FIGURE 6-2. Physical and Biological Routes of Transport of Hazardous Substances, Their Release from Disposal Sites, and Potential for Human Exposure. Source: Grisham, J. W., Ed. 1986.

Clearly, with the production and use of toxic chemicals in this country, we have the potential for prolific contamination of our water supplies. To prevent this from occurring, there has been an attempt within the public sector to identify toxic waste sites, and to characterize the risks they pose to exposed populations (EPA 1984; Harris et al. 1984). Several of the sites have been evaluated to determine their risk to public health, however most of the existing studies have involved single site attempts to look at health outcomes (Clark et al. 1982; Logue et al. 1985; Najem et al. 1985).

Although it has been suggested that it isn't worthwhile to combine data on health status gathered from groups of individuals who have been exposed at different sites (Maugh 1982), there have been a few attempts to evaluate human health risks through multisite studies (Vainio et al. 1981; EPA 1984). The difficulty in using multisite studies can be described quite simply: Almost every site contains a complex mixture of known and unknown compounds. In fact, you can expect that waste materials and exposure settings will be different at each disposal site (Heath 1983). These are critical factors in establishing epidemiologic studies since the epidemiologists must be able to identify and evaluate the toxicity of the chemicals in the site, assess the potential for human exposure and estimate associated biological risks. Along with the problems associated with multisite studies, there are also some pluses.

For example, exposed populations at dump sites are often very small, and therefore an increased number of sites can potentially increase the size of the study population. This will increase the statistical power of the analysis to detect true differences among members of the study group. The increased size of study populations will also make it possible to address the issue of latency concerning the study of chronic disease endpoints: The larger the study population the more man-years of exposure to evaluate.

Ambient Air Contamination

We must also be concerned about the millions of tons of chemical wastes emitted into the ambient air annually (for example, sulfur dioxide [SO_2], nitrogen dioxide [NO_2], respirable suspended particulates [RSP], and ozone [O_3]) that all too frequently exceed the NAAQS of 100 $\mu g/m^3$. Frequently, indoor concentrations of these same pollutants far exceed the outdoor concentrations. Many attempts have been made over the years to study the effects of air pollution on human health.

For example, in December 1952, for a period of four days, London, England, was covered by a dense fog. Although the public realized that the fog appeared to be unusually severe, not much thought was given to the potential health problems. Although health professionals recognized that

112   Environmental Epidemiology

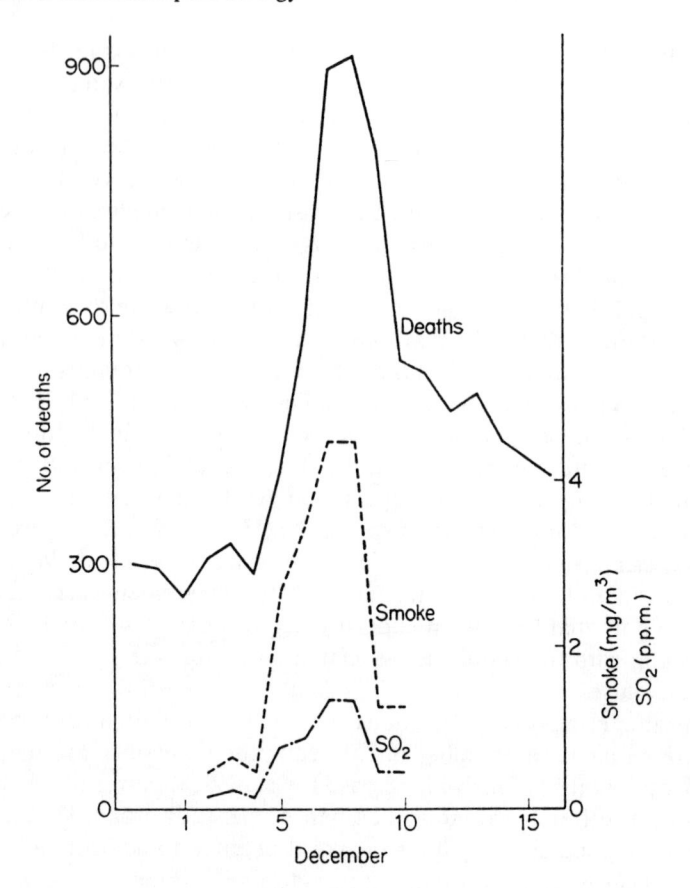

FIGURE 6-3.   Total Deaths in Greater London and Air Pollution Levels Measured During the Fog of December 1952. Source: Alderson 1976.

there was increased demand for hospital beds, it was not until after the fog had disappeared, and health data were analyzed, that the true dimensions of the tragedy were made known to the public: Mortality (4,703 deaths) during the four days from December 5–8 increased 2 1/2 times what it had been for that same period the year before. Although the winter was particularly cold that year, many health professionals believed that the fog also had an important role to play in the excess deaths. In Figure 6-3, it is clear that the number of deaths increased as the atmospheric level of sulfur dioxide and total suspended matter increased. Two years after the fog, a report released by the authorities suggested that the number of deaths from bronchitis during the fog was more than nine times as high as for the preceding week.

Although there is some consideration given to the premise that highly sensitive, or susceptible people react more adversely to chemical exposures, and thus provide a sentinel of warning to the rest of the community, this has not always proven to be the case. For example, Goldsmith and Breslow reported on excess mortality among the elderly during a heat wave in Los Angeles, California, at a time when ozone levels were particularly high. As shown in Figures 6-4 and 6-5, excess deaths among the elderly, including nursing home patients, were related to the high temperatures but not to increased levels of ozone (Goldsmith and Breslow 1959).

Although many of the earlier studies on environmental pollutants failed to capture actual exposure data, it seems clear that chemical discharge into the ambient environment (for example, soil, air, and water) poses potential health risks, largely unmeasured, to a substantial portion of our population. Although acute health symptoms resulting from contact with many of these environmental pollutants is well known, the long-term health effects issue is relatively unexplored.

ENVIRONMENTAL AND BIOLOGICAL MONITORING OF EXPOSURE

Environmental monitoring is designed to demonstrate the concentration of the toxicant in the sampled air, plants, or soil, and to predict the values (that is, concentrations) that will be found in true exposure situations. Chemical residues are difficult to measure, and because they are not uniform in size and may move rapidly over large geographical areas in variable concentrations, their distribution is difficult to predict. Unfortunately, failure to accurately estimate exposure levels may lead to misclassification in epidemiologic studies. The use of biological samples (for example, blood, urine) to monitor the concentrations of chemical residues in tissue and fluid in exposed persons should greatly improve exposure assessment, thereby reducing the likelihood of misclassification. The ability to use biological markers (see Chapter 8) for exposure and effect should greatly enhance the quantification of dose at the target organ site.

There are three major routes of exposure to toxic chemicals (Kilgore and Akesson 1980; and Laughlin and Gold 1988):

1. Respiratory (Inhalation)
2. Dermal
3. Oral

Respiratory Exposure

Air measurements generally involve collecting chemical residues from the air onto a filter, sorbent trap, or combination of the two. The sample is then

114   Environmental Epidemiology

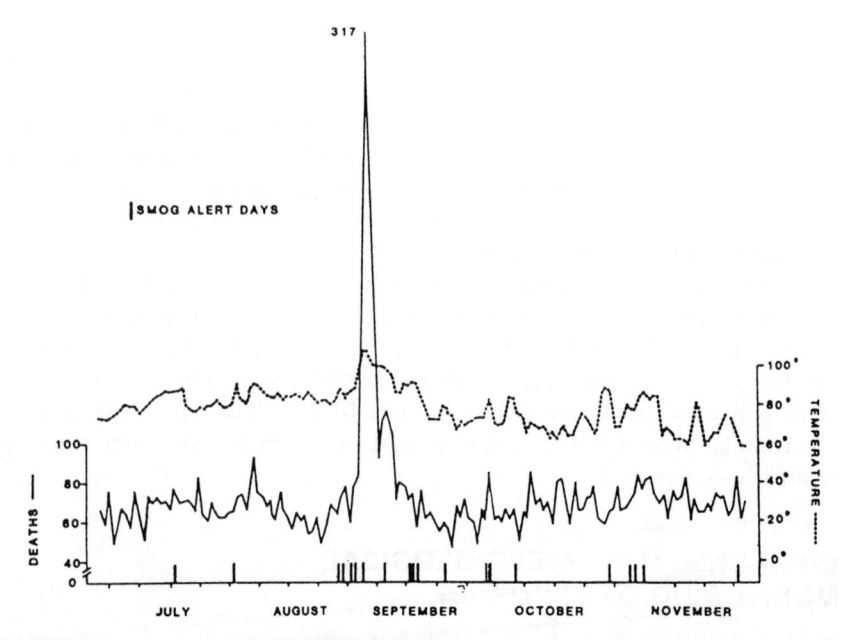

FIGURE 6-4. Association of Mortality of Persons Over 65 Years of Age and Temperature During the Heat Wave of 1955. On the Days Marked Smog, Alerts Occurred and the Ozone Levels Exceeded 0.5 ppm. Source: Goldsmith & Breslow 1959.

extracted and analyzed. Samples may be collected over a 24-hour time period, or for shorter periods, depending on need. The finer the pollutant residue, the greater the proportion of dispersed material that can be expected to be inhaled. Thus, respiratory exposure is greatest for aerosols, intermediate for dusts, and lowest for dilute spray formulations (Wolfe et al. 1967).

Attempts have been made to measure chemical residues in the ambient air using spatial monitors located in fixed positions, as well as in an especially selected microenvironment (for example, an indoor work area, or living area, or an outdoor work or play area). Chemical residues are monitored in an individual's breathing zone by the use of personal air samplers (for example, a small sorbent trap attached to a battery-powered pump weighing about 1 kg.). Data gathered in this manner are then combined with an activity log (questionnaire that records actual sources, times, duration of exposure, and frequency of exposure) that is completed by the individual. **The spatial concentration weighted by the time within the microenvironment provides the estimated exposure**(Davis 1980; Devine et al. 1986; Winterline et al. 1986).

Sampling media will vary depending upon the chemical to be monitored,

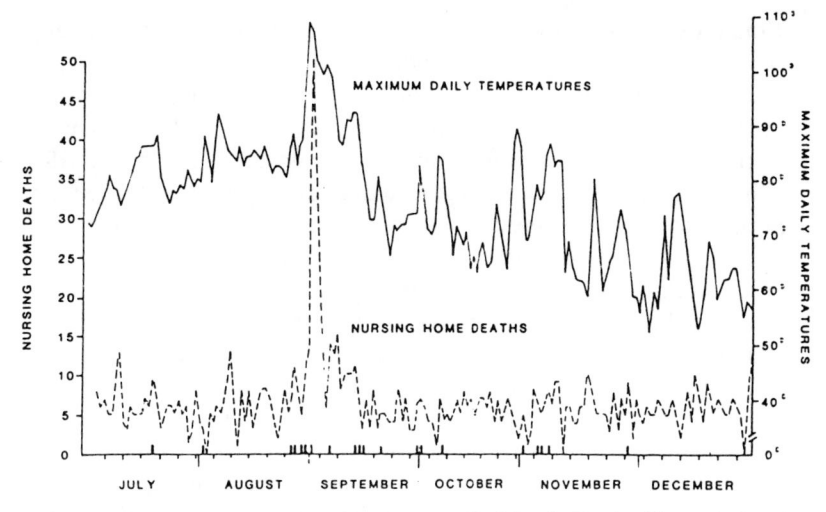

FIGURE 6-5. Association of Mortality of Persons Resident in Nursing Homes in Los Angeles and Heat During the Heat Wave of 1955. Smog Alerts Occurred During the Days Indicated. Note That the Temperature Scale is Different From This Figure, but the Same Temperatures Are Recorded. Source: Goldsmith & Breslow 1959.

although it will probably include polyurethane foam (PUF). Polyurethane foam has enjoyed wide use because of less resistance to air flow than the granular sorbents. However, PUF is not suitable for collecting nonvolatile compounds, such as herbicide salts.

Dermal Exposure

Dermal exposure may be estimated indirectly by measuring residues on clothing, from concentrations in the ambient air, or from absorbent pads (for example, alpha-cellulose for measuring sprays, pads made from thicknesses of surgical gauze and backed by filter paper for measuring chemical dust) attached to parts of the body or clothing of the subject (Durham and Wolfe, 1962). The operating assumption is that the area of the body covered by the pad is representative of the portion of the body being studied. Although the patch technique is widely used in exposure assessment, there are limitations (for example, how much of the exposure will the patch absorb, and how much can be reclaimed for analysis), and alternative techniques to improve upon the accuracy, precision, and representativeness of the sample must be developed with each study. Rather than patches, highly absorbent knit white cotton garments that cover the entire body part being studied (for example, gloves to estimate exposure to the hands, short sleeved undershirts for the upper torso) may provide a more accurate estimate of dermal exposure.

Dermal exposure may also be measured directly from the surface area of the skin of exposed subjects by swabbing an area of exposed skin with surgical gauze sponges saturated with 95 percent ethyl alcohol (Harris et al. 1984). After the swab has been used it is placed in a sealed jar and returned to the laboratory for analysis (investigators have determined that to remove 90 percent or more of the chemical parathion residue from an area of skin the size of the back of a man's hand it is necessary to use several swabs). Hand rinses can also be used to estimate dermal exposure. As described by Durham and Wolfe (1962), the hands of the subject must be clean before exposure. Upon testing, the subject's hand is placed into a plastic container, usually a sturdy bag holding about 200 ml of 95 percent ethyl alcohol or other harmless solvent, and the fingers of the hand are rubbed briskly against the thumb and the palm to remove particles. The hand is removed from the bag, and the bag is tightly bound and sent to the laboratory for analysis. Two hand rinses were found to remove as much as 96 percent of the chemical parathion from one hand when done shortly after exposure.

Although the skin is an important route of exposure to toxicants, it has proven difficult to determine the amount of chemical residues actually reaching the skin's surface through clothing and other protective materials (Milby et al. 1964; Hayes et al. 1964).

Oral Exposure

Oral exposure to a toxic chemical occurs through ingestion of material that is covered by or has absorbed the toxicant (for example, apples sprayed with the pesticide Alar), either by accident or intent. Clearly, a major hazard in the general population today is the threat to children who chew and swallow paint and dirt contaminated by lead residues (for example, flaking lead-based paints from tenement homes; dust and soil in and about homes that are near heavy concentrations of tetraethyl lead exposure from gasoline emissions).

ABSORPTION

Absorption can be defined as the amount of substance absorbed into body tissues and fluids. Absorption may be estimated directly from the measurement of residues or metabolites of chemicals found in body tissues and fluids.

Dermal Absorption/Dosage

Human skin comes in contact with many toxic chemicals. Fortunately, skin is not highly permeable and it serves as a fairly good fatty barrier between man and his environment. However, there are chemicals that can permeate skin tissue and produce systemic effects (for example, organophosphorus (op) insecticides can rapidly permeate the skin, resulting in acute illness, and death). In Figure 6-6 we see a diagram of a cross-section of human skin. To permeate the skin, the toxicant must pass through the epidermal cells, hair follicles, oil (sebaceous) glands, or the sweat glands. For percutaneous (through the skin) absorption to occur, the chemical must travel through the packed layers of horny, keratinized epidermal cells, on through the germinal layer of the epidermis, through the corium (the dermis or true skin), and finally into the systemic circulation.

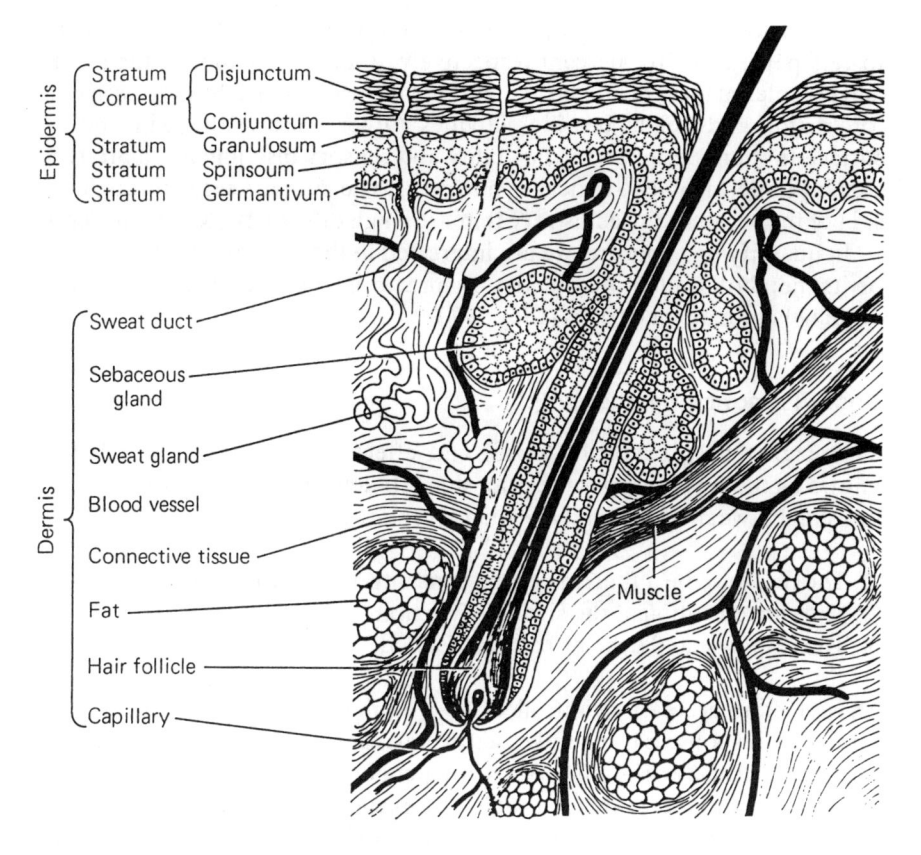

FIGURE 6-6. Diagram of a Cross-Section of Human Skin. Source: Klaassen 1980.

Although we can measure chemical residues with great sensitivity in body tissue (for example, adipose, blood), at levels in the ppt range, it has proven very difficult to **quantify** the amount of chemical residue absorbed through the skin (Ware et al. 1975; Maibach et al. 1971). We do know that when a fat-soluble (lipid) chemical joins with lipid-saturated skin, chemical absorption is promoted (Wester and Maibach 1983). The concentration of dose also plays an important role in absorption. For example, it has been shown that, as the concentration of a toxic chemical to human skin increased from 4 $\mu g/cm^2$ to 2,000 $\mu g/cm^2$, tissue absorption increased from 0.34 $\mu g/cm^2$ to 180.0 $\mu g/cm^2$ respectively (Maibach and Feldmann 1975). Wester and Maibach (1983) have also shown that by increasing the surface area of an applied dose, absorption could be increased.

The anatomic site plays a role in chemical absorption. Maibach and Feldmann (1975) applied a 4 $\mu g/cm^2$ concentration of parathion to the forearm, abdomen, and forehead of volunteer participants and found that absorbed dose (as measured by urine concentration) ranged from 8.6 percent to 18.5 percent to 36.3 percent respectively. In fact, the palms of the hands and the soles of the feet are the least permeable tissue, since they are designed for friction and weight bearing activities. Other areas of the skin's horny surface that are designed for flexibility and sensory discrimination are reasonably impermeable to toxicants; however, toxicants pass easily through the skin of the scrotum since it is very thin with high diffusivity (Klaassen 1980).

In quantifying exposure, we are interested in the amount of the chemical or its metabolite identified at the organ site in the body from which the ensuing health effect derives (Lioy 1990).

Dose-Response Relationship

Characteristically, exposure (dose) and effect come together in what is known as the dose-response relationship. To determine a quantitative relationship between a dose and a response requires that:

1. The response be a function of the concentration of the dose at the site;
2. The concentration at the site be a function of the dose; and
3. The response and dose by causally related.

These factors are easily controlled in experimental settings. However, the term dose-response is frequently used in nonexperimental epidemiology with the finding of an association between a dependent variable (for example, lung cancer) and an independent variable (for example, smoking). Although the use of dose-response provides a sense of credibility when considering causal relationships, the use of the term is misleading and inappropriate in the true

Characterizing Human Exposure   119

toxicological sense, because in nonexperimental epidemiology quite often (1) the identity of the toxic agent is unknown, (2) true dose to which the organism has been exposed is unknown, (3) and the organ site and specificity of response are unknown.

The use of the term dose-response implies that part or all of the response may be due to the environmental insult. For example, in Figure 6-7, we see that the risk of developing lung cancer increases as the number of cigarettes smoked increase. This is the classic use of dose-response in an epidemiologic study. However, it would be more appropriate to use similar, but more accurate terms to describe the association (for example, exposure-response), since we really do not know: (1) which carcinogen in the tobacco smoke is responsible for the lung cancer, (2) how much of the carcinogen reached the target site, and (3) how much of the carcinogen it took to cause the lung

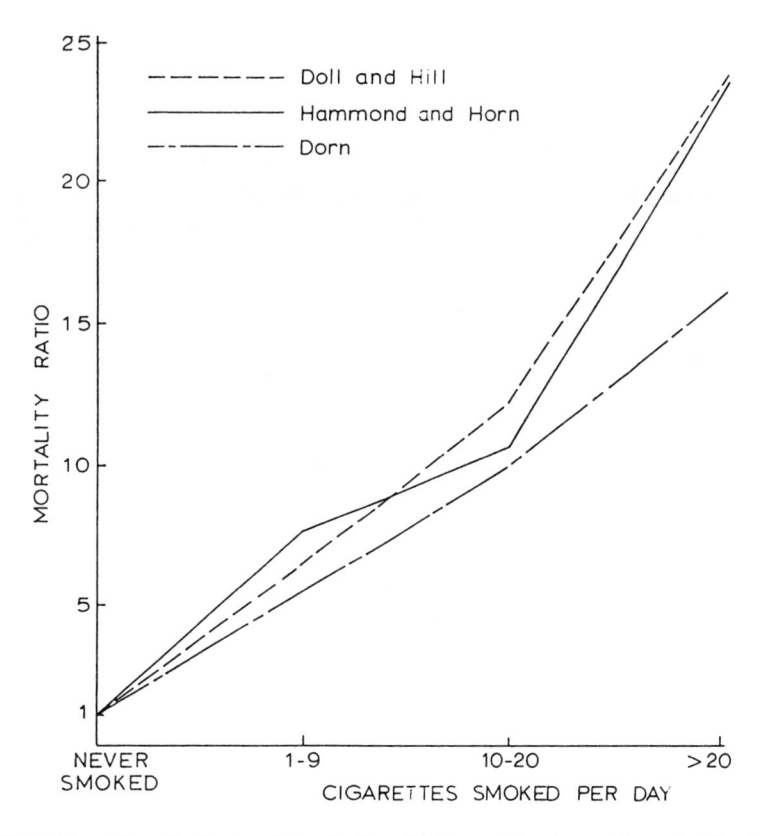

FIGURE 6-7. Mortality Ratios of Deaths From Malignant Neoplasm of Lung by Number of Cigarettes Smoked Daily in Three Prospective Studies. Source: Lilienfeld & Lilienfeld 1980.

cancer. A true dose-response relationship is shown in Figure 6-8. When observing such a response, and applying the three factors listed at the begining of this section, we can make a strong case for causality.

Biological and Biochemical Markers

In the past, although epidemiologists have used surrogates to determine exposure and estimate dosage (for example, numbers of cigarettes smoked per day, distance from a smelter, days worked in a specific work environment), it has always been difficult to reliably associate exposure and absorption levels to adverse health effects. It has been particularly difficult for environmental epidemiologists to associate low-level exposures over a long time period to adverse health effects.

To estimate absorption, one may measure chemical residues in the fluids and tissues of exposed persons (for example, chlorinated hydrocarbon and chlorophenoxy residues in the blood, and OP metabolites in the urine). The amounts of sample required, the time of collection, and the compound to be detected are determined by the nature of the compound of interest. Chemicals that degrade or are metabolized rapidly may be absent in a particular sample, but their original presence may be determined by an analysis of metabolite (for example, alkyl phosphates from OPs, phenols from chlorophenoxy acid herbicides or carbamates, or DDA from DDT).

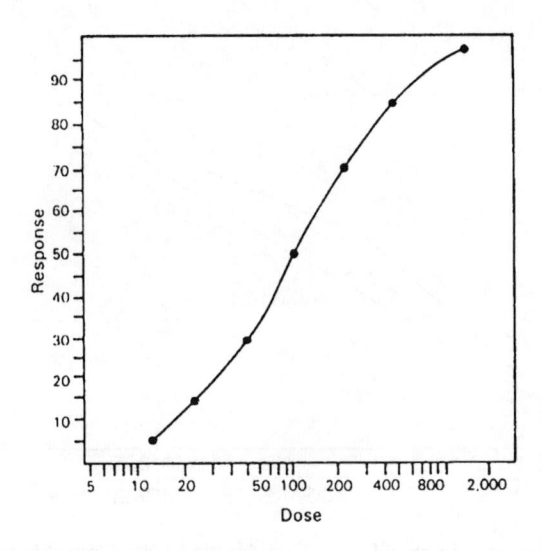

FIGURE 6-8. Diagram of a Dose-Response Relationship. Source: Klassen & Doull 1980.

If body tissues are analyzed quickly following exposure, the likelihood of a positive analysis is enhanced. However, if exposure is low and/or experienced over a long time period, it will likely require metabolite analysis (for example, urinary metabolites from OP insecticides will be found four to eight hours after exposure). Also, when toxic concentrations are thought to be small, larger samples must be taken (for example, morning urine samples, or 24-hour pooled samples).

Recently, the use of biological markers (for example, sister chromatid exchange [SCEs], chromosome aberrations, protein/DNA adducts, mutagens in body fluids, micronuclei, sperm morphology, and somatic cell mutation) has been touted as potentially the most effective means for measuring markers of exposure and/or effect (Vainio et al. 1983). Once a chemical is absorbed into tissue, the focus is on the amount of the chemical substance that has interacted with critical cellular targets or surrogates (**the biologically effective dose**). Available indicators of biologically effective dose reflect interaction with genetic material (nucleic acid or the chromosomes) or surrogate macromolecules (protein) (Perera and Weinstein 1982). Unfortunately, in using markers as surrogates for exposure, specificity is often lacking since the marker may not be specific for a particular chemical. In fact, the use of biological markers requires expensive and sophisticated assay techniques that vary in sensitivity and specificity from population to population. However, when used as markers of exposure, biological markers certainly enhance the ability of the epidemiologist to confirm individual exposure (Griffith et al. 1989).

For example, it is possible to use the blood enzyme cholinesterase (red blood cell and plasma cholinesterase) as a marker of exposure to OP and carbamate chemicals. However, the measurement of whole blood cholinesterase (ChE) activity as a marker would require the establishment of baseline ChE values on workers prior to collection. Generally, the use of biological markers is expensive because of the high costs of assays and the need to develop appropriate tests for sensitivity and specificity. Thus, study populations are often quite small (see Chapter 8 for a more detailed discussion of biochemical and biological markers).

Environmental Monitoring

Environmental monitoring requires the development of detailed protocols that allow chemical residues of known integrity and sufficient quantity to be collected. Protocols should be structured to facilitate the use of the bioassay, fractionation procedures and chemical analyses necessary to identify the relationships between the sample, and the source environment (that is, ambient and microenvironments). The protocol should address techniques required to collect particulate material as well as volatile, semivolatile, and condensable organic material. Protocols should also provide information on methods related to source, ambient, and microenvironment collection activities. Typical activities for sources will include sampling with extractive sampling equipment and appropriate particulate filtration and vapor phase organic collection. Ambient sampling would be expected to include size separation of particulate matter, the collection of respirable and inhalable particulates, and other organic materials. Particle size fractionation would be included as appropriate.

A Case Study in Ambient Monitoring

In Zenica and Lukavac Yugoslavia, approximately 50 steel-producing coke ovens have been in operation for about 30 years. Since it has been reported that the highest concentrations of poly aromatic hydrocarbons (PAHs), including BaP, are topside on the coke oven battery (Hemminki et al. 1990; IARC 1988), the fine particle sampler shown in Figure 6-9 was placed on a catwalk approximately 1 1/2 meters above the oven lids on top of the battery.

During the eight-hour sampling period, the test apparatus was exposed to visible emissions of effluent from the coking process. The sampling instrument was located to most effectively collect emissions resulting from the charging and coking process (Keimig et al. 1986). The sampler was especially designed to collect duplicate fine particle ($< 2.5 \mu m$ in aerodynamic diameter) samples. A cyclone inlet preceded the filter pack to remove particles > 2.5 μm. One fine particle sample was collected on a 47 mm diameter quartz filter. Simultaneously, a duplicate sample was collected on a 47 mm diameter Teflon filter. The mass and elemental composition was determined on the sample collected on the Teflon filter.

Fine particle mass concentration measured during the eight-hour shift was 1.871 and 1.458 mg/m^3 at Lukavac and Zenica respectively. These are extremely high values when one considers the NAAQS for PM-10 aerosols for a 24-hour period is 150 $\mu g/m^3$. At both coke oven facilities sulfur (probably in the form of sulfate) and Chlorine were the highest inorganic species in the fine particle samples. The Gas Chromatograph/Mass Spectrometer (GC/MS) analysis was performed on an aliquot of a CH_2Cl_2 extract of these quartz filters. One half of each filter was weighed and Soxhlet extracted with 200 ml of methylene chloride for 47 hours. Internal standards (5.0 μg each of perdeuterated naphthalene, phenanthrene, and chrysene) were added to the extracts and the volumes were reduced to less than 5 ml on a rotary evaporator. The extracts were transferred to concentrator tubes, blown down to 5 ml under nitrogen, and transferred to Teflon-capped vials for storage.

The GC/MS analyses were performed on a HP 5970 MSD with a 5890 GC

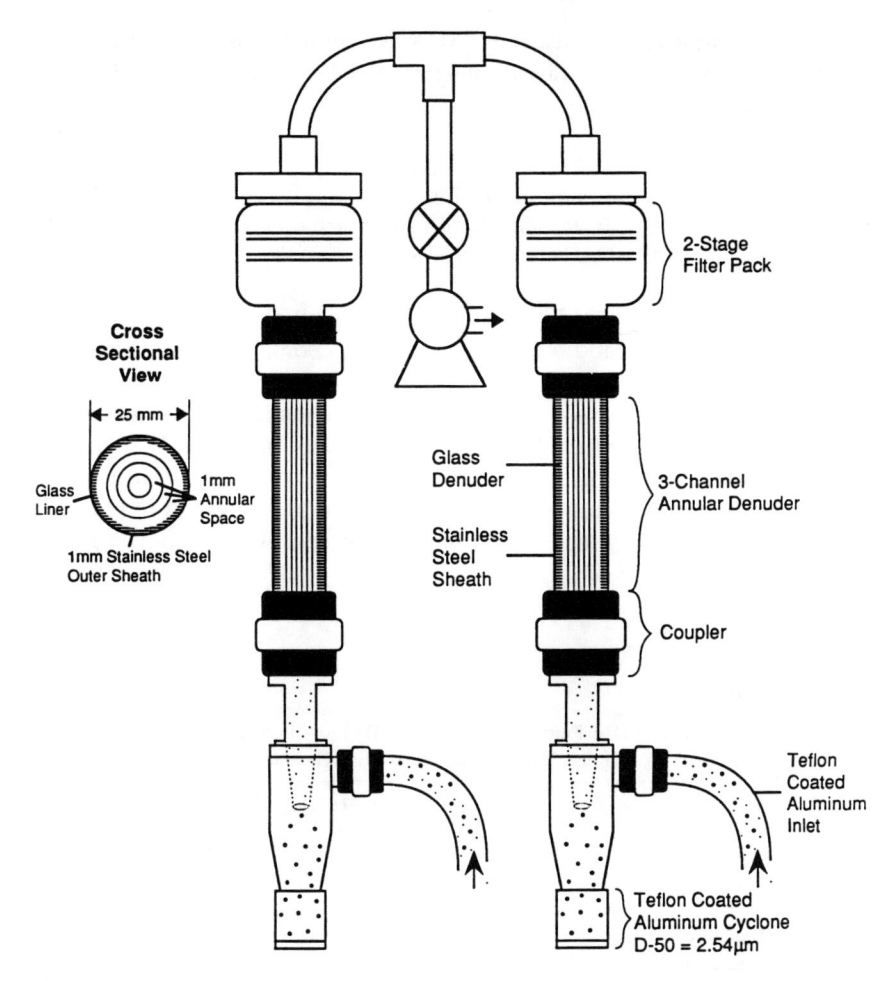

FIGURE 6-9. Fine Particle Sampler. Source: By permission, Griffith et al. 1991.

and capillary direct interface. The GC oven temperature was held at 47°C for 1 minute, programmed at 15°C/minute to 200°C, then at 3°C/minute to 310°C, where it was held for 10 minutes. The GC/MS interface temperature was 260°C. The GC column was a 25m × 0.2mm HP Ultra-2 with a 0.11μm film and a 1m × 0.32mm retention gap for the qualitative analyses. The quantitative analyses were performed with a 30m × 0.25mm DB-5 column, 0.25μm film, without a retention gap. In both cases, the carrier gas was helium at about cm/s and cold on-column sample injection was used. The MSD was operated under autotune conditions for all analyses, with an electron multiplier voltage of 1,600 v.

For the qualitative analyses, the MSD was scanned from 50 to 500 amu at 0.95 s/scan. For the quantitative analyses, the MSD was operated in the selected ion-monitoring mode with the following program:

5 - 12 min- m/z 128, 136, 152, 154 for 75ms each
12 - 20 min- m/z 166, 178, 188, 202 for 75ms each
20 - 44 min- m/z 228, 240, 252, 276, 278 for 75ms each
44 - 58 min- m/z 300, 302, 326 for 100ms each

Quantitative analyses of PAHs and trace element compositions are summarized in Tables 6-1 and 6-2 respectively, and compared with literature values reported in other studies. Poly aromatic hydrocarbon emissions in the Yugoslavian plant in Lukavac, when matched by chemical and compared by the sign test to the plant in Zenica, are statistically significantly higher ($p < 0.01$). The plants in Zenica and Lukavac appear to emit significantly more ($p < 0.05$; $p < 0.001$) PAHs, respectively, compared to the plant in Norway. This is so whether nondetected are treated as zero, or whether only positive values are used.

TABLE 6-1. Distribution of Selected PAHs Identified Topside in Zenica, Lukavac, and Oslo Coke Ovens.

	PAHs ($\mu g/m^3$)		
	Zenica	Lukavac	Oslo*
Naphthalene	nd	.2	nd
Acenaphthylene	nd	6.8	nd
Acenaphthene	nd	1.3	nd
Fluorene	.6	31.1	nd
Phenanthrene	15.0	163.4	3.0
Anthracene	3.8	63.8	1.3
Fluoranthene	32.7	163.4	8.0
Pyrene	24.5	116.7	7.1
Benz[a]anthracene	20.4	93.4	5.8
Chrysene	20.4	85.6	7.0
Total Benzofluoranthenes	27.3	124.5	4.3
Benzo[a]pyrene	15.0	60.7	7.3
Indeno[1,2,3 − c + d]pyrene	9.0	36.6	4.5
Dibenz[ah]anthracene	4.2	18.7	1.2
Benzo[ghi]perylene	7.8	33.5	4.4
Sum selected PAHs	180.7	999.6	53.9

* Haugen et al. (1986).

TABLE 6-2. Fine Particle Mass and Elemental Composition.

Species	Concentration ($\mu g/m^3$) Zenica	Lukavac
Mass	1871.000	1458.000
Si	0.902	1.007
S	13.315	15.623
Cl	52.661	46.927
K	0.384	0.410
Ca	2.349	0.758
Fe	0.589	2.096
Zn	0.436	0.109
Br	0.794	0.563
I	0.546	----
Pb	0.387	0.138
Ni	ND	0.027
Cu	ND	0.041

SUMMARY

The potential adverse health outcomes resulting from involuntary exposure to dangerous substances in the environment has touched health professionals and citizens alike. The media visibility of tragic environmental episodes has often led to a crisis with regard to the responsiveness of health agencies to public concern, and the expectation of effective and immediate remediation. At the heart of the dilemma for environmental epidemiologists is a clear understanding of the issues involved in characterizing human exposure. In this chapter, we have attempted to touch on all the major issues involved in exposure assessment, ranging from identifying the pollutant to estimating dosage. Exposure assessment is the single most important facet of environmental epidemiology, for without an accurate estimate of exposure, we will be unable to associate environmental exposures to adverse health outcomes.

ASSIGNMENTS

1. Define exposure characterization and describe the ten steps involved in the process.
2. What are the three major routes of exposure?
3. Define dose-response. Name the three factors that are required before a dose-response relationship can be determined.

GLOSSARY

Absorption Barrier: Any of the exchange barriers of the body that allow differential diffusion of various substances across a boundary. Examples of absorption barriers are the skin, lung tissue, and gastrointestinal wall.

Absorbed Dose: The amount of a substance penetrating across an absorption barrier (the exchange boundaries) of an organism, via either physical or biological processes.

Absorption Barrier: Any of the exchange barriers of the body that allow differential diffusion of various substances across a boundary. Examples of absorption barriers are the skin, lung, tissue, and gastrointestinal wall.

Bioavailability: The state of being capable of being absorbed and available to interact with the metabolic processes of an individual. Bioavailability is typically a function of chemical properties, physical state of the material to which an organism is exposed, and the ability of the individual to physiologically absorb the chemical or agent.

Dose-Responsive Curve: A graphical representation of the quantitative relationship between administered, applied, or internal dose of a chemical or agent, and a specific biological response to that chemical or agent.

Dose-Response Relationship: The resulting biological responses in an organ or organism expressed as a function of a series of different doses.

Dose: The amount of a substance available for interaction with metabolic processes or biologically significant receptors after crossing the outer boundary of a person. The potential dose is the amount ingested, inhaled, or applied to the skin. The applied dose is the amount of a substance presented to an absorption barrier and available for absorption. The absorbed dose is the amount crossing a specific absorption barrier (for example, the exchange boundaries of the skin, lung, and digestive tract).

Dosimetry: Process of measuring dose.

Environmental Fate: The fate of a chemical or biological pollutant following release into the environment. Environmental fate involves temporal and spatial considerations of transport, transfer, storage, and transformation.

Exposure: Contact of a chemical, physical, or biological agent with the outer boundary of an individual. Exposure is quantified as the concentration of the agent in the medium integrated over the duration of the exposure.

Exposure Assessment: The determination or estimation (qualitative or quantitative) of the magnitude, frequency, duration, and route of exposure.

Exposure Scenario: A set of facts, assumptions, and inferences about how exposure takes place that aids the exposure assessor in evaluating, estimating, or quantifying exposures.

Fixed Location Monitoring: Sampling of an environmental or ambient medium for pollutant concentrations at one location continuously or repeatedly over a period of time.

High End Exposure or Dose Measurement: A plausible estimate of the individual risk for those persons at the upper end of the risk distribution. Conceptually placed above the 90th percentile, but not higher than the person in the population with the highest exposure or dose.

Internal Dose: The amount of a substance penetrating across the absorption barriers (the exchange boundaries) of a person, via either physical or biological processes.

Method Detection Limit (MOD), or the Limit of Detection (LOD): The minimum concentration of an analyte that, in a given matrix (a specific medium, for example, surface water, drinking water) and with a specific method, has a 99 percent probability of being identified, quantitatively or qualitatively measured, and reported to be greater than zero.

Microenvironment: Well-defined surroundings such as the home, office, automobile, kitchen, store, etcetera, that can be treated as homogeneous (or well characterized) in the concentrations of a chemical or other agent.

Microenvironment Method: A method used in predictive exposure assessments to estimate exposures by sequentially assessing exposure for a series of areas (microenvironments) that can be approximated by constant or well-characterized concentrations of a chemical or other agent.

Pathway: The physical course a chemical or pollutant takes from the source to the exposed person.

Personal Measurement: A measurement collected from an individual's immediate environment using active or passive devices to collect the samples.

Point of Contact Measurement: An approach to quantifying exposure by taking measurements of concentration over time at or near the point of contact between the chemical and a person while the exposure is taking place.

Potential Dose: The amount of a chemical contained in material ingested, breathed, or applied to the skin.

Route of Exposure: The way a chemical or pollutant enters a person after contact (for example, ingestion, inhalation, or dermal absorption).

REFERENCES

Alderson, M. 1976. *An Introduction to Epidemiology.* PSG Publishing Co., Inc. Massachusetts.

Berlin, M. 1985. "Low level Benzene Exposure in Sweden: Effect on Blood Elements and Body Burden of Benzene." *American Journal of Industrial Medicine* 7:365–73. A smoking history was gathered on each worker.

Brookes, P., Lawley, P. 1964. "Evidence for the Binding of Polynuclear Aromatic Hydrocarbons to Nucleic Acids of Mouse Skin: Relationship Between Carcinogenic Power of Hydrocarbons and Their Binding to DNA." *Nature* 202:781–84.

Bjørseth, A., Bjørseth, O., Fjeldsted, E. 1978. "Polycyclic Aromatic Hydrocarbons in the Work Atmosphere." *Scand. J. Work Environ. & Health* 4:224–36.

Cohen, D. B., and Bowes, G. W. *Water Quality and Pesticides: A California Risk Assessment Program, Vol. 1.* State Water Resources Control Board Report No. 84-65P. Sacramento, Calif.: State Water Resources Control Board.

Davies, J. 1980. "Minimizing Occupational Exposure to Pesticides: Personnel Monitoring." *Residue Reviews* 75:33–50.

De Meo, M.P., Dumnil, G., Botta, A.H., Laget, M., Zabalouef, V. Mathias A. 1987. "Urine Mutagenicity of Steel Workers Exposed to Coke Oven Emissions." *Carcinogenesis* 8:363–67.

Devine, J. M., Kinoshita, G. B., Peterson, R. P., Picard, G. L. 1986. "Farm Worker Exposure to Terbufos [phosphorodithioic acid, S-(tert-butylthio) methyl O,O-diethyl ester] During Planting Operations of Corn." *Archives of Environmental Contamination and Toxicology* 15:113–19.

Durham, W., and Wolfe, H. R. 1962. "Measurements of the Exposure of Workers to Pesticides." *Bull WHO* 26:75–91.

EPA (U.S. Environmental Protection Agency, Office of Policy Planning and Evaluation). 1988. *Environmental Progress and Challenges: EPA's update.* EPA-230-07. 88–033. Washington, D.C.: U.S. Government Printing Office.

EPA. National Priorities List. 1984. *786 Current and Proposed Sites in Order of Ranking and by State, October, 1984.* HW-7.2. Revised, December.

Everson, R. B., Randerath, E., Avitts, T. A. et al. 1987. "Preliminary Investigations of Tissue Specificity, Species Specificity, and Strategies for Identifying Chemicals Causing DNA Adducts in Human Placenta." *Prog Exp Tumor Res* 31:86–103.

Gelboin, H. V., and Ts'O.P.O.P. 1978. *Polycyclic Hydrocarbons and Cancer, Vol. 1 and 2.* Academic Press, Inc., New York.

Goldsmith, J. R., and Breslow, L. 1959. "Epidemiological Aspects of Air Pollution." *J. Air Pollut. Cont. Assoc.*

Goldstein, B. D. 1988. "The Scientific Basis for Policy Decisions." In *Epidemiology and Health Risk Assessment.* Ed. L. Gordis. New York: Oxford University Press.

Griffith, J., Duncan, R. C., and Hulka, B. 1989. "Biochemical and Biological Markers: Implications for Epidemiologic Studies." *Archives of Environmental Health* 44:375–81.

Griffith, J., Stevens, R. K., Duncan, R. C., Everson, R. B., Whiton, R. S., Gallagher, J. E., Lewtas, J. E., Schramm, M., and Cherkez, F. Submitted to: *Archives of Environmental Health.*

Grisham, J. W., ed. 1986. *Health Aspects of the Disposal of Waste Chemicals.* New York: Pergamon Press.

Gupta, R. C., Reddy, M. V., Randerath K. 1982. "32P-Postlabeling Analysis of Non-Radioactive Aromatic Carcinogen-DNA-Adducts." *Carcinogenesis* 3:1081–92.

Harris, R. H, Highland, J. H., Humphreys, K. 1984. "Comparative Risk Assessment for Remedial Action Planning." *Hazardous Waste* 1:19–33.

Harvey R. 1982. "Polycyclic Hydrocarbons and Cancer." *American Scientist* 70:386–93.

Haugen A, Becher G, Benestad C, Vahakangas K, Trivers G. E., Newman M. J., Harris C. C. 1986. "Determination of Polycyclic Aromatic Hydrocarbons in the Urine, Benzo(a)Pyrene Diol Epoxide-DNA Adducts in Lymphocyte DNA, and Antibodies to the Adducts in Sera From Coke Oven Workers Exposed to Measured Amounts of Polycyclic Aromatic Hydrocarbons in the Work Atmosphere." *Cancer Research* 46:4178–83.

Hayes, G. R., Jr., Funckes, A. J., Hartwell, W. V. 1964. "Dermal Exposure of Human Volunteers to Parathion." *Archives of Environmental Health* 8:829.

Hemminki, K., Grzybowska, E., Chorazy, M., Twardowska-Saucha, K., Sroczynski J. W., Putman K. L., Randerath K, Phillips D. H., Hewer A. 1990. "Aromatic DNA Adducts in White Blood Cells of Coke Workers." *Int Archives of Occupational and Environmental Health* 62:467–70.

Heath, C. W., Jr. 1983. "Field Epidemiologic Studies of Populations Exposed to Waste Dumps." *Environmental Health Perspectives* 48:3–7.
HMSO. Health and Safety Executive. *Benzene.* (Toxicity Review No. 4.) London: 1982.
IARC. International Agency for Research on Cancer. 1988. *Methods for Detecting DNA Damaging Agents in Humans: Applications in Cancer Epidemiology and Prevention.* IARC Scientific Publication No. 89, IARC, Lyon.
IARC. International Agency for Research on Cancer. 1982. "Monographs on the Evaluation of Carcinogenic Risks of Chemicals to Humans." *Some Industrial Chemicals and Dyestuffs.* IARC Scientific Publication No. 29, IARC, Lyon.
Jackson, J. O., Warner, P. O., Mooney, T. F. 1974. "Profiles of Benzo(a)pyrene and Coal Tar Pitch Volatiles at and in the Immediate Vicinity of a Coke Oven Battery." *American Industrial Hygiene Assocociation Journal* 35: 276–81.
Keimig, D. G., Slymen, D. J., and White, O. Jr. 1986. "Occupational Exposure to Coke Oven Emissions From 1979–1983." *Archives of Environmental Health* 41:363–67.
Kilgore, W. W., and Akesson, N. B. 1980. "Minimizing Occupational Exposure to Pesticides: Populations at Exposure Risk." *Residue Review* 75:21–31.
Klaassen, C. D. 1980. "Absorption, Distribution, and Excretion of Toxicants." In Cassarett and Doull's *Toxicology 2nd Ed,* ed John Doull, Curtis Klaassen, and Mary Amdur. New York: Macmillan Publishing Co., Inc.
Klaassen, C. D., and Doull, J. 1980. "Evaluation of Safety: Toxicologic Evaluation." In Cassarett and Doull's *Toxicology 2nd Ed,* ed John Doull, Curtis Klaassen, and Mary Amdur. New York: Macmillan Publishing Co., Inc.
Kreyberg L. 1959. "3:4 Benzpyrene in Industrial Air Pollution: Some Reflections." *British Journal of Cancer* 13:618–22.
Lagakos, S. W., Wessen, B. J., Zelen, M. 1986. "An Analysis of Contaminated Well Water and Health Effects in Woburn, Massachusetts." *Journal of the American Statistical Association* 81:583–96.
Laughlin, J., Gold, R. 1988. "Cleaning Protective Apparel to Reduce Pesticide Exposure." In *Reviews of Environmental Contamination and Toxicology* 101:93–119.
Lawther, P. J., Commins, B. T., Waller, R. E. 1965. "A Study of the Concentration of Polycyclic Aromatic Hydrocarbons in Gas Works Retort Houses." *British Journal of Industrial Medicine* 22:13–20.
Lindstedt G, Sollenberg J. 1982. "Polycyclic Aromatic Hydrocarbons in the Occupational Work Environment." *Scand J Work Environ Health* 8:1–19.
Lilienfeld, A. M., and Lilienfeld, D. E. 1980. *Foundations of Epidemiology.* New York: Oxford University Press.
Logue, J. N., Stroman, R. M., Reid, D., Hayes, C. W., Sivarajah, K. 1985. "Investigation of Potential Health Effects Associated with Well Water Chemical Contamination in Londonderry Township, Pennsylvania, U.S.A." *Archives of Environmental Health* 40:155–60.
MacMahon B. Comment. 1986. *Journal of the American Statistics Assocciation* 81: 597–99.
Maibach, H. I., and Feldmann, R. J. 1975. "Systemic Absorption of Pesticides Through the Skin of Man." In *Occupational Exposure to Pesticides: Report to the Federal Working Group on Pest Management from the Task Group on Occupational Exposure to Pesticides, Appendix B,* pp. 120–27. U.S. Government Printing Office (GPO), 0-551-026, Washington, D.C.

Maibach, H. I., Feldmann, R. J., Milby, T. W., Serat, W. F. 1971. "Regional Variation in Percutaneous Penetration in Man." *Archives of Environmental Health* 23:208–11.

Masek, V. 1971. "Benzo(a)pyrene in the Workplace Atmosphere of Coal and Pitch Coking Plants." *Journal of Occupational Medicine* 13:193–98.

Maugh, T. H. 1982. "Just How Hazardous Are Dumps?" *Science*215:490–93.

Milby, T. H., Ottoboni, F., and Mitchell, H.W. 1964. "Parathion Residue Poisoning Among Orchard Workers." *JAMA* 189:351–56.

NAS. National Academy of Science. *Particulate Polycyclic Organic Matter.* Washington, D.C., 1972.

National Research Council. *Environmental Epidemiology: Public Health and Hazardous Wastes.* National Academy Press, Washington, D.C., 1991.

Najem, G. R, Louria, D. B., Lavenhar, M. A., Feuerman, M. 1985. "Clusters of Cancer Mortality in New Jersey Municipalities; With Special Reference to Chemical Toxic Waste Disposal Sites and Per Capita Income." *Journal of Epidemiology* 14:528–37.

NIOSH. Code of Federal Regulations Title 29, Part 1910, 1976. U.S. Dept. of Labor, Occupational Safety and Health Administration. *Final Occupational Safety and Health Standard for Exposure to Coke Oven Emissions.* Fed Reg 41:46742–93.

OTA (U.S. Congress, Office of Technology Assessment). 1989. *Coming Clean: Superfund Problems Can Be Solved.* OTA-ITE-433. Washington, D.C.: U.S. Government Printing Office. 223

Paigen, B., and Goldman, L. R. 1987. "Lessons from Love Canal, New York, U.S.A.: The Role of the Public and the Use of Birth Weights, Growth, and Indigenous Wildlife to Evaluate Health Risk." pp. 177-192 in *Health Effects from Hazardous Waste Sites.* J. B. Andelman and D. W. Underhill, eds. Chelsea, Mich.

Perera, F., and Weinstein, I. B. 1982. "Molecular Epidemiology and Carcinogen-DNA Adduct Detection: New Approaches to Studies of Human Cancer Causation." *Journal of Chronic Disease* 35:581–600.

Perera, F. P., Santella, R. M., Brenner, D., Poirier, M. C., Munshi, A. A., Fischman, H. K., Van Ryzin, J. 1987. "DNA Adducts, Protein Adducts and Sister Chromatid Exchange in Cigarette Smokers and Nonsmokers." *Journal of the National Cancer Institute* 87:449–56.

Perera, F. P., Hemminki, K., Young, T. L., Brenner, D., Kelly, G., Santella, R. M. 1988. "Detection of Polycyclic Aromatic Hydrocarbon-DNA Adducts in White Blood Cells of Foundry Workers." *Cancer Research* 48:2288–91.

Phillips, D. H, Hewer, A., Grover, P. L. 1986. "Aromatic DNA Adducts in Human Bone Marrow and Peripheral Blood Leucocytes." *Carcinogenesis* 7:2071–75.

Philips, D., and Sims P. 1979. "PAH Metabolites: Their Reaction With Nucleic Acids." In: *Chemical Carcinogens and DNA, Vol. 2* (P. L. Grover, Ed.), Boca Raton, FL: CRC Press, Inc. pp. 9–57.

Phillips, D. H., Hemminki, K., Alhonen, A., Hewer, A., Grover P. L. 1988. "Monitoring Occupational Exposure to Carcinogens: Detection by 32P-Postlabeling of Aromatic DNA Adducts in White Blood Cells From Iron Foundry Workers." *Mutation Research* 204:531–41.

Reddy, M. V., Randerath, K. 1986. "Nuclease P1-Mediated Enhancement of Sensitivity of 32P-Postlabeling Test for Structurally Diverse DNA Adducts." *Carcinogenesis* 7:1543–51.

Schulte, K. A., Larsen, D. J., Hornung, R. W., Crable, J. V. 1975. "Analytical Methods Used in a Study of Coke Oven Effluent." *American Industrial Hygiene Association Journal* 131–39.

Sexton, K., Letz, R., and Spengler, J. D. 1983. "Estimating Human Exposure to Nitrogen Dioxide: An Indoor/Outdoor Modeling Approach." *Environmental Research* 32:151–66.

Sheehy, J. W. 1980. *Control Technology for Worker Exposure to Coke Oven Emissions.* U.S. Dept. Health, Education, and Welfare, NIOSH Pub. No. 80-114.

Tanimura H. 1968. "Benzo(a)pyrene in an Iron and Steel Works." *Archives of Environmental Health* 1968;17:172–77.

Tennant, R. W., Margolin, B. H., Shelby, M. D., Zeiger, E., Haseman, J. K., Spalding, J., Caspary, W., Resnick, M., Stasiewica, S., Anderson, B., Minor, R. 1987. "Prediction of Chemical Carcinogenicity in Rodents From *in vitro* Genetic Toxicity Assays." *Science* 236:933–41.

Thielen, R. G. 1979. "Benzo(a)pyren-Konzentrationen bei Arbeiten auf Koksbatterien." *Jahresber Dtsch Ges Arbeitsmed*:267–74.

van Schooten, F. J., van Leeuwen, F. E., Hillebrand, M. J. X., de Rijke M. E., Hart, A. A. M., van Veen, H. G., Oosterink, S., Kriek, E. 1990. "Determination of Benzo(a)pyrene Diol Epoxide-DNA Adducts in White Blood Cell DNA From Coke-Oven Workers: The impact of Smoking." *Journal of the National Cancer Institute* 82:927–33.

Vainio, H., Sorsa, M., Hemminki, K. 1983. "Biological Monitoring in Surveillance of Exposure of Genotoxicants." *American Journal of Ind Medicine* 4:87–103.

Vainio, H., Sorsa, M., Rantanen, J., Hemminki, K., Aitio A. 1981. "Biological Monitoring in the Identification of the Cancer Risk of Individuals Exposed to Chemical Carcinogens." *Scand J Work Environ Health* 7:241–51.

Ware, G. W., Morgan, D. P., Estesen, B. J., Cahill, W. P. 1975. "Establishment of Reentry Intervals for Organophosphate-Treated Cotton Fields Based on Human Data:III. 12 to 72 Hours Post-Treatment Exposure to Monocrotophos, Ethyl-and Methyl Parathion." *Archives of Environmental Contamination and Toxicology* 3:289–306.

Weinstein, I. B. 1978. "Current Concepts on Mechanism of Chemical Carcinogenesis." *Bulletin of the N.Y. Academy of Medicine* 54:336–83.

Wester, R. C., and Maibach, H. I. 1983. "Cutaneous Pharmacokinetics: 10 Steps to Percutaneous Absorption." *Drug Metabolism Review* 14:169–205.

Winterline, W. L., Kilgore, W. W., Mourer, C. R., Hall, G., Hodapp, D. 1986. "Worker Reentry into Captan-Treated Grape Fields in California." *Archives of Environmental Contamination and Toxicology* 15:301–11.

Wogan, G. N., Gorelick, N. J. 1985. "An Overview of Chemical and Biochemical Dosimetry of Exposure to Genotoxic Chemicals." *Environmental Health Perspectives* 62:5–18.

Wolfe, H. R., Durham, W. F., and Armstrong, J. F. 1967. "Exposure of Workers to Pesticides." *Archives of Environmental Health* 14:622–33.

7
Laboratory Practice

Vincent F. Garry, Jack Griffith, and Tim E. Aldrich

OBJECTIVE

This chapter will describe the laboratory procedures essential to epidemiologic studies of human environmental disease.

INTRODUCTION

In environmental epidemiology, interest focuses directly on the interaction between exposure to an agent and the reaction of the human organism to that exposure. At this interface, the role of the laboratory is to identify and define the pathobiology of exposure and its relationship to the disease endpoint under consideration. Toward this effort, laboratory investigation may be either broad-based or narrowly defined. Laboratory activity may vary from analysis of a specific chemical to a laboratory-based investigation of the cause(s) of a malfunction of the organism due to agent exposure.

LABORATORY PRACTICE
General Guidelines

Test systems developed and used by the laboratory in environmental epidemiologic studies are measures of biologic, chemical, or physical properties of the material world. The extent to which these measures accurately represent the property under study is limited by the degree to which theoretical considerations match observation. For these reasons, a test system should contain an **internal control standard** and an **independent external control**.

For analytic chemical or physical determinations, internal or machine controls are usually part of the instrumentation. For biologic or biochemical analyses, **internal standards are usually derived standards**. Purified materials with known activities or concentrations from a nationally recognized source are examples of a derived internal standard. An external standard may consist of pooled materials from a number of specimens repeatedly examined over time or specimens repeatedly drawn from a standard control population obtained over time. Because of inherent variability of biologic activity, biologically based test systems require an external control.

The requirement for internal controls other than machine controls for biologic test systems is fraught with difficulty because of the imprecision of many biologic substrates. In these biologic test systems there are additional requirements for positive (that is, controls that display the biologic effect to be studied), and negative (that is, controls that do not contain the biologic effect) controls. All of these efforts are essential internal efforts of the laboratory to assure reproducibility of a test system. Whether the laboratory test is assessed as a true measure of a chemical, physical, or biologic property of the material under study is dependent on relationships to reference technologies, biologic outcome, and continued confirmation of the theoretical basis of the test system.

Materials for Analysis

Air, water, soil, body fluids, and tissues are the primary sources of materials for laboratory analysis and interpretation. In this chapter, we will place emphasis on **direct measures of human exposure (dosage) and measures of human pathobiologic effects obtained from materials derived from body fluids and tissues.**

Studies Involving Human Subjects

Clearly defined approval by a legitimate human-subjects review board is a *sine qua non* for conducting laboratory analyses for epidemiologic purposes. Subjects must be fully informed and understand the purpose of the laboratory study, the risks involved, and the availability of laboratory derived information to them as a result of their participation. Signed informed consent must be obtained from all subjects prior to testing (Figure 7-1).

Specimen Collection

Specimens submitted for laboratory examination must be defined and labeled as to source and date of collection. The manner in which the specimen was collected and prepared, method of transport, and times of collection and

I hereby give my consent for my participation in the project Entitled: <u>An Assessment of Field Worker Occupational Exposure to Pesticides in Florida Agriculture.</u>
I understand that the person responsible for this project is Dr. _____.
He (She) has explained that these studies are part of a project that has the following objectives (Briefly explain objectives) <u>To investigate the use of pesticides in the citrus industry, farmworkers exposure to these pesticides, and their possible health effects.</u>

He (She) and his (her) authorized representative has (1) explained the procedures to be followed and identified those which are experimental; (2) described the attendant discomforts and risks; (3) described the benefits to be expected, if any; (4) described appropriate alternative procedures; and (5) agreed to answer any questions I may have concerning the procedure(s), a description of which are as follows: (Briefly and in simple terms state procedures; specifically identifying those which are experimental) <u>Participation in this study requires each person to answer a questionaire about his/her job, pesticides used, and his/her state of health. Also, each person on some farms would be asked to permit researchers to collect urine and blood samples for analysis.</u>

The risks have been explained to me as follows: (List all risks of more than negligible probability and/or severity) <u>Risks are those associated with donating a blood sample; including: slight pain or hematoma in the area of the puncture, possible infection, or irritation and bleeding.</u>

I understand that in the event of physical injury resulting from the research procedures described to me, the University of Miami, and their affiliates are not able to offer financial compensation nor to absorb the cost of medical treatment. However, necessary facilities, emergency treatment and professional services will be availble to research subjects, just as they are to the community generally. Further information about any of the above matters may be obtained from the University of Miami, School of Medicine, 15655 S.W. 127 Ave., Miami, FL 33177, Phone (A/C 305) 251-9823.

The duration of time of my participation covered by this consent form is: _____ season; however, I understand that I may discontinue at any time I choose and my withdrawal will not affect my future treatment at this institution.

_____
Signature of Subject   Date

_____
Signature of Parent/Guardian or Authorized Representative (if required)

_____
Signature of Project Director or his Authorized Representative

_____
Signature of Witness to Oral Presentation and Signature

FIGURE 7-1. Sample Informed Consent Form from a University of Miami Study.

receipt in the laboratory should be recorded. Specimens for chemical analysis should be prepared in chemically clean containers. Specimens for biologic analysis should be aseptically prepared and placed in containers free of microbiologic and chemical contaminants. In the preparation and transport of materials for biologic and chemical analysis, care must be taken to protect the sample from the possible effects of light (especially ultraviolet (UV) light) and temperature on the stability of the specimen. Extreme temperature variation either in transport or storage is to be avoided. X-ray examination of the specimens during shipping is a potential problem for biologic as well as chemical analysis of the specimen.

Specific Requirements for Specimen Collection and Preparation Media

Air: Temperature, humidity, wind velocity and direction, the collection device and its placement, the volume to be collected, and the sorbent material for the specimen to be collected are all primary concerns when monitoring for air pollutants. Daytime hours versus nighttime hours of collection are important variables for semivolatile gases and UV-sensitive contaminants. In the occupational setting, precise placement of personal monitoring devices and area monitoring devices are added concerns. Extraneous contaminants and failure to maintain the integrity of the specimen are common sources of error in reporting data from this media.

Water: Temperature, depth, location, flow rate, volume and direction of flow, and season are important variables in collecting water specimens. Chemically or biologically contaminated containers are often sources of error in data collection from this media.

Soil: The type of soil, depth, moisture content, temperature of the soil, location, season, and the types of microorganisms present in the soil are all considerations in collecting specimens for laboratory determinations. Collection errors commonly arise in the methods and collection devices used (for example, bore samples inadequately maintained) at any point in the collection process.

HUMAN TISSUES AND BODY FLUIDS

Tissues: Method of preservation, sample size and segment of tissue selected for analysis, ambient temperature, rate of decomposition, and presence of infectious agents are all specific factors to be considered in collecting human tissues for analysis. Common collection errors occur in sample handling. For example, analysis for trace elements can be confounded by the use of nickel/chrome alloy surgical instruments in the collection of the specimen.

Formalin preserved tissues may interfere with analysis of organic chemical contaminants and biologically relevant macromolecules including DNA (that is, human genetic material, deoxyribonucleic acid, contained within subcellular structures called chromosomes constitute the "identity" of the living cell).

Blood: Blood and blood components (serum, plasma, red blood cells, white blood cells, and platelets) are the most easily accessed human tissue. Phlebotomy technique, methods of preparation, preservation and transport, presence of infectious agents (in particular, hepatitis and AIDS viral agents) are variables to be considered in any blood collection scheme. The most common collection errors center on the phlebotomy technique and method of blood preservation. With regard to phlebotomy, inadvertent trauma in blood collection can induce hemolysis of red blood cells and alter the coagulation properties of the specimen. This is a primary source for laboratory artifact in determining serum enzyme activities, viability, and proliferative capacity of white cell subsets. Source or shelf-life (variation in the potency of the specific preservative used by batch lot), is another major source of blood collection error that will adversely affect laboratory determinations.

Model Blood Drawing Protocol for an Epidemiologic Field Study

Supplies for blood draws and finger sticks are as follows:

Water source
Lancets
Drinking cups
Hypodermic needles (nontoxic, nonpyrogenic, nonreactive to tissue. Size 20 or 21 Gx1 or 11/2 inches.)
5 ml serum separator tubes
10 ml sodium heparin tubes
Subject ID labels
Glass slides
Aluminum foil
Gauze pads
Individually wrapped isopropyl alcohol (70%) swabs.
Sterile Band-Aid plastic strips
Paper towels
Fruit juice
Blood sample data forms

Hand soap
Blood collection kits
Destruclip hypodermic safety device
Sterile disposable needles
Cleansing towels
Betadine
Tourniquets
Test tube racks

The phlebotomist: The phlebotomist (M.D., R.N., Licensed Practical Nurse, Licensed Laboratory Technician, or Certified Phlebotomist) will be responsible for setting up the blood drawing site, maintaining supplies, and explaining procedures to study participants.

Blood collection site: The blood collection site is set up for the safety and convenience of the study participants. However, every effort should be made to ensure that the following equipment is available in close proximity to the site: lavatory, sink, and refrigerator. Bloods should be drawn, if possible, where there is good light, and a cot should be provided for any participant feeling discomfort and needing to lie down. Additional work space should be available for supplies when a subject prefers to lie down for the phlebotomy.

Drawing blood: The phlebotomist should be at ease and in no apparent rush. Blood samples will be collected in vacutainer tubes, placed in a test tube rack, and kept at room temperature not exceeding 85° for 30–45 minutes to allow for plasma separation.

Sample Labels:

> Blood
> Full Name
> Date and Time Collected
> Participant ID

The samples will then be placed in a styrofoam container with wet ice and kept in this manner until the day's sampling is completed. Upon returning to a headquarters location, the samples will be prepared for shipping to the laboratory when the phlebotomist has concluded separation procedures.

Supplies should be conveniently located with ease of accessibility. The subject should be made comfortable in a sitting position and encouraged to relax. The subject I.D. should be confirmed prior to drawing blood. As the phlebotomist prepares for the blood draw he or she should explain that a sample of blood will be taken from the tip of the finger and from the arm. Do not draw blood from any area with open sores or wounds, rashes, or that appears to be swollen.

The phlebotomist should:

1. Wash hands thoroughly.
2. Wear latex gloves at all times.
3. If at all possible draw blood from the antecubital area of the arm. If an appropriate vein cannot be found, draw blood from a hand, using a small (23g) needle.
4. The tourniquet should be applied about 3–4 inches above the elbow. Ask the subject to hold his or her arm down and to open and close hand, making a fist, several times.
5. Palpate veins, and select one that is not hard or has no visible scarring.
6. A 19 g or 21 g collection set should be used on all normal veins. A 23 g collection set should be used for difficult draws. Obese persons will likely need a 20 g 1 1/2 inch needle.
7. Do not maintain the tourniquet in place for more than one minute at a time before the blood draw.
8. Cleanse puncture site with alcohol wipe. Dry the area with 2×2 gauze. Using the "butterfly" needle, with the bevel pointed up, approach the vein at a 15° angle in the direction of the vein approximately 1/2" to 3/4" below the point of entry.
9. Fix vein by pulling the skin tight.
10. Use the micropore tape to tie the "butterfly wings" together after the needle is through the skin, and blood has appeared in the tubing of the butterfly.
11. Puncture the stopper of the vacutainer tube by placing the tube (at the center of the stopper) on the butterfly needle. It may be necessary to reposition the needle if no blood appears in the tube. It may even be necessary to remove the needle and try again on the other arm.
12. The blood in the tube must not be allowed to touch the stopper.
13. The tourniquet should be removed when the first tube is about half full. Let the subject relax. When the tube is full, replace it with the second tube. Do the same with other tubes. Invert each tube several times, but do not shake, and place the tubes on the blood tube rocker.
14. The serum separator tube should be allowed to clot at room temperature for 30–45 minutes.
15. If at any time during the blood draw, any damage to the subject occurs (for example, a hematoma) terminate the draw and move to the other arm.

Processing Blood Specimens

The blood will be spun down by centrifugation for 10 minutes at approximately 3,000 rpms. At the conclusion of every blood draw work, surfaces should be cleaned with a solution of bleach and water (1:5) or other suitable disinfectant. Plastic-lined drape sheets should be used to protect all supplies.

Urine

Urine collection practices are highly varied and dependent on the quantitative or qualitative laboratory endpoint under consideration. Common collection errors arise out of a lack of understanding of the laboratory endpoint to be studied. For example, quantitative analysis for toxicant metabolites may be affected by the presence of actively metabolizing bacteria in a urine specimen collected 24 hours or more prior to analysis. Precautions such as asepsis in collection, refrigeration of the specimen, and use of a preservative can avoid this collection error. Urine is a final common pathway for excretion and concentration of many toxicants, therefore diluting the specimen is not acceptable. Precautions such as measuring specific gravity of the urine, measuring serum and urinary creatinine, and where necessary, direct visual observation of the collection procedure can avoid this collection error. Because of the difficulties inherent in 24-hour specimen collections, first voided morning urine specimens after toxicant exposure may be used to provide an estimate of toxicant metabolite concentration. Similarly, in the occupational setting, sequential urine specimens obtained at the beginning and end of the work shift or work week can provide an estimate of toxicant metabolite concentration.

To summarize, collecting urine specimens for environmental and occupational epidemiologic purposes requires careful attention to the laboratory endpoint to be examined and to study subject compliance.

Model Urine Collection Protocol for an Epidemiologic Field Study

Participants are asked to urinate in washed and hexane-rinsed laboratory specimen bottles. Each bottle should contain a blank label securely attached so as not to come off during shipment or transportation in dry ice (Avery S-2441 labels 1" × 3" size).

Sample Labels:

> *Urine*
> Full Name
> Date and time collected
> Participant ID

Bottle caps should be lined with Teflon or aluminum foil. Each participant should be handed a specimen bottle and at that time the subjects's name should be written indelibly on the bottle label. After the subject returns the specimen bottle with urine, it should be immediately placed in dry ice in a styrofoam container so the sample will be immediately frozen. All participant samples collected at the study site should be transported immediately in dry ice by the most efficient means of transportation to the laboratory.

When packaging the samples for shipment, they should be placed in an insulated container with dry ice (about 2 lbs of dry ice per urine sample). The insulated container should be packed in a strong cardboard box to protect the insulated container from damage. The outside of the package should be marked with the number of pounds of dry ice in the container, and the letters **ORM-A** written on the box within a rectangle (without any of the letters touching the sides of the rectangle) to identify the package as medical samples. The transporter should be told that the package contains samples for medical research purposes (this will probably permit the samples to be shipped without the normally required shipping certificates). Shipping should be by overnight freight to ensure that samples arrive at the laboratory before thawing.

Sample Custody and Processing for Blood and Urine Samples

Samples sent from the field via courier should be received by a representative of the laboratory staff. A laboratory aide should assign an identification number to each sample and enter the number into a permanent, bound record. This record should record whether the sample was serum or urine, the name of the subject, the individual identification number, the site of the research location, the date and time the sample was collected, the date and time the sample was received at the laboratory, the name of the responsible researcher, and any miscellaneous remarks. Columns should be provided for data analysis and values for each residue analyzed.

For chemical residue analysis, serum samples should be maintained in a frozen state at -12 to $-180C$ pending analysis. Storage should **never** be in a freezer containing standard solutions (for fear of cross-contamination). When the time for analysis arrives, the sample should be defrosted, a 2-cc aliquot removed with a clean pipette or syringe, and the stock solution quickly refrozen and stored at −12 to −180C. For analysis of genetic markers, lymphocyte culture should be undertaken as soon as the sample is taken to the laboratory. Although heparinized blood can be stored for several days at 4–370C, it is possible that attempts to store samples may lead to the death of the lymphocytes, reduction of mitogenic response, and selective loss of cells with lesions that can result in SCE formation (Bloom 1981). An aliquot of every nth (15th, 25th, etcetera) sample should be made available for quality control analysis.

Urine samples should be defrosted. A 15 cc aliquot of the shaken urine should be poured into a centrifuge tube, centrifuged, and used to determine creatinine and osmolality values. Two 20 cc aliquots should also be placed in prelabeled, clean vials and refrozen pending analysis. Storage of the frozen urine should be at −12 to −180C, and never around standards. An aliquot of every nth (10th, 15th, 25th) urine should be made available for quality control analysis.

LABORATORY QUALITY CONTROL

Every effort should be made to ensure that a quality control program is in place. A sampling procedure should ensure that aliquots from a selected number of samples be sent to a cooperating laboratory, or otherwise independent system, for confirmation of analysis. Procedures should be in place to assess data precision, accuracy and completeness (e.g., spiked urine samples should be taken through the entire procedure as should be the reagents to determine the presence of any interfering component). Each set of samples analyzed on any given day should contain at least one reagent blank and one spiked (deliberately positive) sample.

Hair and Nails

Hair and nails are easily accessed sources of biomaterials for analysis. Some trace metal contaminants bioaccumulate (that is, arsenic) in these tissues. Some organic chemical contaminants can be identified in hair and nails. Pathologic variations in the protein composition of these tissues are detectable. Because the growth rate of these tissues over a period of months **can be determined**, segmental analysis of portions of the hair shaft and nails can define exposure incidents occurring several months before tissue sampling. Standardizing methods for collecting and analyzing materials from this important biologic source has lagged. Sources of collection error include use of chemically contaminated instruments and containers. Hair tints and dyes, shampoos, and nail polish are especially noxious sources of inadvertent chemical contamination.

Seminal Fluid and Sperm

Current laboratory methodologies used to detect reproductive hazards in humans are fraught with medical, legal, and ethical overtones. At the technical level, seminal fluid collection for many routine procedures is complicated by the need to maintain viable sperm. Viability is exquisitely sensitive to temperature, trace chemical contaminants, and time from collection to assay. Of these factors, time from collection to assay is the most critical to field studies and is a common source of collection error. The World Health Organization recommends that no more than two hours elapse from the time of collection, until the assay is begun.

Respired Air

Metabolic status, levels of volatile toxicants and their metabolites can be directly measured in respired air. Because of the limitations imposed by the complex technology involved in these assessments, field use is limited. Stability of the sample is the chief limitation for specimens collected in the field.

SPECIFICITY AND SENSITIVITY OF LABORATORY ASSAY METHODS

To characterize any laboratory assay and assess the relative usefulness of any laboratory-based parameter, two factors are considered:

1. The specificity of the method, and
2. Sensitivity of the assay (Griffith et al. 1989).

Specificity is the ability of the assay to detect a unique event to the exclusion of all other events (that is, to what extent can the assay detect a specific chemical or a specific pathobiologic effect to the exclusion of all other similar chemicals or similar pathobiologic effects). Sensitivity refers to the ability of an assay to detect unique events at a level statistically significant from some background level (that is, it is the ability of the assay to pick out a signal from biologic or analytic noise at some concentration).

Typically, the quantitative laboratory data is compared to some range of normal or to some detection limit. It is the laboratory's responsibility to assure that set detection limits will eliminate recording extraneous signals from similar materials (that is, that the limit of sensitivity of the assay does not exceed the level of specificity). It is also the laboratory's responsibility to ensure that deviations from a normal range measure some property of one or more pathobiologic events related to the occurrence of disease. The degree to which the event is indicative of disease, organ, or system dysfunction at a given concentration of the analyte (substance or material analyzed) is a measure of specificity and sensitivity of the assay.

Biologic Dose

Biologic dose is a twofold measure. First, it is a measure of concentration of an analyte; second, it is a measure of a therapeutic or noxious biologic effect. In the laboratory, individual methods are routinely devised to assess one singular measure of dose. The search for laboratory methods that will directly measure concentration and biologic effect in the same assay is a continuing laboratory effort. Current methodology, at best, provides indirect assessment of biologic dose either in terms of concentration or biologic effect. In the following paragraphs we will discuss the two modes of measurement of dose.

Toxicant concentration: Determination of toxicant concentration commonly involves extraction and concentration of the toxicant from the specimen under study and finally quantitative analysis of the desired analyte (Silverstein and Bassler 1968; Baselt 1980). The most critical features of this laboratory approach are the efficiency of the extraction method to acquire and concentrate the desired analyte, and the specificity and sensitivity of the analytic method. Analytic methods designed to quantitatively measure toxicant concentration tend to rely on specific spectra of depicting the structure of the chemical analyte or specific elution of the chemical in a solvent system. Gas chromatography (GC) and high pressure liquid chromatography (HPLC) take advantage of the quantitative separation of analytes achieved in a solid or solvent matrix to measure concentration. Ultraviolet, infrared, atomic absorption, and mass spectroscopy identify and quantify structurally unique characteristics of a chemical analyte in the presence of a radiation source of given wavelength. The specificity of these methods is dependent on the uniqueness of (1) the separation achieved in a given media for chromatography and (2) the structural characteristics of the chemical given a radiation source for spectroscopic methods. Sensitivity of these methods is dependent on the amount of analyte needed to detect differences from background.

Biomarkers of Toxicant Concentration

Few biologically based or biochemically based assays are designed to assess toxicant concentration. These assays are dependent on specific quantitative binding of a chemical to well-defined macromolecular sites on DNA or protein. For example, polyaromatic hydrocarbons form detectable covalent linkage (adducts) to specific DNA and protein sites. These adducts are quantifiable by immunologic, HPLC (high pressure liquid chromatography), and radiometric methods (Gupta and Earley 1988; Perera et al. 1988; Santella et al. 1990). The efficacy of this laboratory approach is limited by the ability to detect specific structural alteration and binding of biologic macromolecules by toxicant chemicals.

Model Method for Adduct Analysis in a Field Study

The ^{32}P-Postlabeling assay (Gupta and Earley 1988) presents considerable promise in the investigation of DNA adducts. For example, the sensitivity of ^{32}P-Postlabeling is able to detect a lone adduct from an aromatic hydrocarbon per cell, or as little as one microgram (μ) of adduct per 100 kg of tissue. Assays can be performed with as little as 10 μg of DNA, which is the product of about 10 mg of tissue. Peripheral white blood cells will be isolated and fractionated into specific cell (granulocytes, lymphocytes, and monocytes) by density gradient techniques. Although present methods are more efficient at detecting relatively polyaromatic hydrocarbon adducts, they are sufficiently sensitive that they can detect several adducts of known and unknown structure.

Since erythrocytes (red blood cells) have no nucleus, and thus no chromosomes, analysis of adducts in these cells is limited to protein adducts of hemoglobin. However, since hemoglobin is relatively easy to obtain, and since the life span of an erythrocyte is long (120 days) relative to other cells, it is possible to investigate cumulative dosage in chronic exposure studies using red blood cells (RBCs). Although hemoglobin adducts are not believed to play a direct role in tumor induction, they may effectively serve as a surrogate for other processes. For example, in humans, small alkylating agents, such as ethylene oxide, are uniformly dispersed throughout the body, and the amount of hemoglobin alkylation generally approximates the dose to DNA (Ehrenberg et al. 1983).

Biomarkers of Dose

Enzymes: Xenobiotic chemical concentration, dependent enhancement, or inhibition of specific enzyme activity related to a known pathobiologic effect is an ideal measure of dose. To illustrate, the enzyme acetylcholinesterase regulates neurotransmission of nerve impulses. The extent of disruption of normal neurotransmission by chemical inhibitors of cholinesterase enzyme activity is directly related to concentration of the enzyme in the system. Similarly, the extent and severity of neuropathophysiologic effects of inhibition of the enzyme is related to the concentration of the chemical toxicant. Chemical inhibitors of cholinesterase enzyme (for example, organophosphate and carbamate pesticides) in general compete with acetylcholine, the neurotransmitter substance, for active sites on the cholinesterase enzyme.

Competitive inhibition of cholinesterase activity is based on the structural specificity of the chemical for these active sites on the enzyme (Fukuto 1990). **Since the extent of enzyme inhibition is quantitatively related to concentration and pathobiologic effect, the enzyme marker is a measure of toxicant dose.** And, because of the requirement for a chemical structure that is specific for inhibition, the assay is considered to have specificity for a particular class of chemicals. The sensitivity of the assay is dependent on the method used and the extent of variation in cholinesterase enzyme activity in the population under study (Silk et al. 1979; Maroni et al. 1986).

Specific enzyme markers of acute toxic dose related to specific chemicals or chemical classes is an established mode of laboratory evaluation. Enzyme markers of dose for longer term health sequelae specifically related to chemical exposure are few. The specific enhancement of activity of the P_{450} group of liver enzymes due to past exposure to polychlorinated biphenyls is a beginning effort to explore this possibility (Okey et al. 1986; Jones et al. 1985; Omenn and Seboin 1984; Manz et al. 1991).

Biomarkers of Pathobiologic Effects

Genetic markers: Markers of genetic damage fall into two general methodological categories: cellular and molecular. The majority of cellular methods used for human studies require culture of human lymphocytes (for example, the relatively easy availability of lymphocytes make them very attractive cell populations for cytogenic analysis). Since the life span of a lymphocyte may vary from only a few days to as much as 20 years (Leavell and Thorup 1976), they may be useful for estimating the effects of long-term chemical exposures, although DNA repair occurs quite rapidly. It has also been suggested that since lymphocytes are in contact with many body tissues, it is possible that they may provide an integrated measurement of exposure (Perera et al. 1987).

Culture conditions and duration of culture are important factors in assessing samples for genetic markers. At the molecular level, DNA methods require isolates from cryopreserved or freshly obtained cells or tissues. Purity and integrity of the DNA isolate, the kind of molecular probe used, and inadvertent DNA contamination from extraneous sources are critical features of these assays. For the most part these laboratory methods are sensitive biologic measures of dose. The assays are highly specific with regard to the pathobiologic endpoints measured. Point mutation, chromosome aberrations, and SCEs observable at the light microscopic level and their molecular counterparts are examples of different quantitative endpoints measured by these methods.

Chemicals may express genotoxicity in any one or more of these assay systems by different biochemical mechanisms. Although the majority of known carcinogens are genotoxic, there remain a number of carcinogens that do not express genotoxcity (Butterworth 1990; Nemali et al. 1989). Preliminary data regarding the predictive value of chromosome aberrations (Reuterwall and Albertini 1990) suggests that worker cohorts with highly elevated chromosome aberration frequencies may be at increased risk for cancer. Similarly, persons with chromosome instability syndromes of genetic origin are at much higher risk for specific cancers (Boder 1985; McKinnon 1987) than the general public. Aside from this inferential data, the long-term predictive value of many of these laboratory tests from the viewpoint of human cancer has not been made.

Model Laboratory Assay for SCE

The SCE assay involves staining one member of a chromatid pair in replicating cells differently from its sister chromatid. This differential staining helps researchers detect genetic material exchanged between sister chromatids. The assay is based on the incorporation of 5-bromodeoxyuridine (BrdUrd), thymidine analog, into replicating chromosomes so that newly formed SCEs can be seen.

The SCE assay using peripheral blood is a simple and reproducible means of quantifying mutagen exposure, and provides a relatively inexpensive procedure for analyzing samples gathered during field investigations. Since SCE measurements are affected by chemical exposures, it is critical that careful occupational, home, and medical histories be developed to characterize exposure, in order that potential confounding can be addressed.

Model Laboratory Assay for Chromosomal Aberrations

Within six hours of taking the blood sample, whole blood (0.5) should be added to a lymphocyte medium (4.5 ml) that contains media (RPMI 1640), 20 percent fetal calf serum (Reheis, Armour Pharmaceutical Co.), Phytohemaglutinin (0.2 ml Pha-m, GIBCO Co.) and Penicillin-Streptomycin solution (1,000 units/culture). Four cultures should be established per individual. Each culture should be coded and the identity of the donor coded into a laboratory notebook. The whole blood cultures should then be incubated for 48 hours at 37°C. Two hours before harvest, colchicine should be added to arrest the lymphocytes undergoing mitosis. Cells should be treated with a warm hypotonic potassium chloride (0.075 M KCL) solution for eight minutes. Following hypotonic treatment, the cells should be fixed with 3:1 methanol and acetic acid. After fixation, slides should be prepared by a flame dry method and subsequently stained with 2 percent Giemsa stain (Gurr Co.) in phosphate buffer (pH 6.8) for ten minutes. Slides should be analyzed for 100 clearly stained metaphase figures randomly selected from each of the two cultures. In total, 200 cells should be scored for the presence of abnormal chromosome figures using the 1,000 × objective. All abnormal cells should be recorded on a cytogenetic worksheet. Normal variants of chromosome structure should also be recorded (that is, gaps). A typical normal variant, and all abnormal cells, should be photographed using a photomicroscope for permanent record and for analysis by a control reader. The abnormal cells should be classified according to Hirschorn and Cohen (1968). A laboratory control in addition to an on-site control blood specimen should be cultured and analyzed under identical experimental conditions. Results should be expressed as the percentage of complex aberrations per 100 mitoses, microscopic slides, and additional fixed material should be maintained in the laboratory for one year after the study is completed.

Metabolic Markers

These biomarkers are designed to determine the rate, amount, and type of metabolic transformation of xenobiotic chemicals in the host organism. Indigenous enzyme concentration, genetic modification of enzyme structure, quantitative, and qualitative estimates of metabolites of xenobiotics formed in the organism are the chief methodologies used in this laboratory approach. The primary objective in some of these laboratory assessments is the demonstration of enhanced genetic risk for disease or intoxication. One laboratory marker of enhanced genetic risk for bladder cancer is the rate of biotransformation of arylamine xenobiotics by the enzyme N-acetyl transferase (Cartwright et al. 1982) (see Chapter 8 for additional discussion). In another vein, persons whose red blood cells are deficient in the enzyme glucose-6-phosphatase dehydrogenase are at increased risk for intoxication by oxidant xenobiotics (Doull 1980).

Reproductive Markers

There are very few pathobiologic markers for human reproductive effects useful for evaluation of environmental or occupational health issues. In the male, sperm counts, sperm morphology and viability are the primary laboratory methods employed (WHO 1987) as measures of effects of toxicant exposure on male fertility. In the female, markers for pregnancy abnormalities, for example levels of alpha fetoprotein (Brock and Sutcliffe 1972; Taplin et al. 1988; Lemonnier et al. 1990) and amniotic fluid analysis for chromosome or molecular genetic abnormalities (Grouchy and Turleau 1984; Lerman 1986) have not been routinely used to assess effects of workplace reproductive toxicant exposure effects. Measures of female fertility including the use of hormonal biomarkers of female fertility (Wilson 1977) have not been routinely employed in the occupational and environmental setting to measure toxicant exposure effects. Whether these biomarkers for female reproductive status will prove useful in the occupational and environmental setting has not been assessed.

Model Laboratory Approach for Sperm Count

Each individual should be asked to produce a complete medical history, with specific information provided on reproductive history. Semen samples will be collected through masturbation and examined for volume and for sperm density (millions of spermatozoa per ml.) and motility. Measurements of serum testosterone, follicle stimulating hormone (FSH), luteinizing hormone (LH), should be measured by sensitive and specific radioimmunoassays. The FSH and LH values are expressed in International Units (IU)/ml. Testosterone levels are expressed in ng/dl.

Specimen Validity

Earlier parts of this chapter were concerned with specimen collection, laboratory assessment of environmental/occupational toxicant exposure, and measurement of pathobiologic effects. The validity of the specimen is dependent on the quality control measures employed at every point from specimen collection to reporting of data. Specimens should be given a unique identity, labeled, adequately described, handled with maintenance of a chain of custody, and processed in the laboratory with attention to specimen integrity and reported clearly. Results should be described indicating possible sources of error. Detailed records of specimens and results should be maintained in secured long-term storage. Access to these records should be defined beforehand. Finally, retrieval procedures to reacquire these data should be established. All of the concerns affect specimen validity.

Quality Assurance

Quality assurance will follow *good laboratory practice* guidelines. This assures maximum database stability, experimental protocol compliance, and adequate quality control procedures within the study laboratory. Particular attributes include:

1. Sample coding by the quality assurance officer to blind the data analysis.
2. Control samples processed with each experimental sample set to document normal baseline responses.
3. Media control lot testing for sterility assurance.
4. Periodic evaluation of protocol compliance by the quality assurance officer throughout the course of experiments.
5. Review of raw data storage and record keeping by the quality assurance officer.
6. Equipment calibration and function monitoring documentation.

Informed Consent

All procedures involved in any environmental epidemiologic study involving human subjects should provide an opportunity for subjects to understand the objective and the study, and to give informed consent (Fig. 7.1). Upon request, subjects should be told of test results, and of the overall study results. All information, including laboratory findings and medical histories, should be maintained within the requirements of patient confidentiality.

Technicians handling blood specimens should also follow appropriate federal guidelines for preventing HIV virus transmission in health care settings. Briefly, all specimens should be handled as if they are infective. Specimens should be transported in containers with secure lids to prevent leaking. Technicians should wear gloves and gown while handling specimens. Laboratory work surfaces should be decontaminated with a chemical germicide after spills and when laboratory work is completed. Contaminated instruments and equipment should be decontaminated, and specimens should be disposed of under an approved policy for disposal of infective waste.

SUMMARY

This chapter is designed to give the student of environmental epidemiology an overview of the methods, practices, and approach of the laboratorian in investigating environmental/occupational health issues. When you evaluate the relative merits of a biological and biochemical marker for use in epidemiologic studies, at least four *desiderata* concerning assays should be considered: The accuracy of a single determination as compared with a known standard; the reproducibility of a series of determinations when done by the same technician, by a group of technicians in the same laboratory, and by different laboratories; the speed from the point of view of availability in case of emergency; and the simplicity of the procedure from the point of view of economy.

Environmental pathology, the laboratory study of human environmental disease, is a rapidly evolving investigative discipline to serve the needs of epidemiologists and other health practitioners to identify and prevent human environmental disease.

REFERENCES

Baselt, R. C. 1980. *Biological Monitoring Methods for Industrial Chemicals.* Davis, CA: Biomedical Publications. pp. 1–301.

Bloom, A. D., 1981. *Guidelines for Studies of Human Populations Exposed to Mutagenic and Reproductive Hazards.* March of Dimes Birth Defects Foundation.

Boder, E. 1985. "Ataxia-telangiectasia: An Overview." In: *Ataxia- aTelangiectasia.* R. A. Gatti and M. Swift, eds. Genetics, a Neuropathology and Immunology of a Degenerative Disease of Childhood. New York: Alan R. Liss, Inc. pp. 1-63.

Brock, D. J. H., and Sutcliffe, R. G. 1972. "Alpha-fetoprotein in the Aantenatal Diagnosis of Anencephaly and Spina Bifida. *Lancet* aii:923–24.

Butterworth, B. E. 1990. "Consideration of Both Genotoxic and Nongenotoxic Mechanisms in Predicting Carcinogenic Potential." *Mutation Research* 239:117–32.

Cartwright, R. A., Rogers, H. J., Hall, D., Glashan, R. W., Ahmad, R. A., Higgins, E., and Kahn, M. A. 1982. "Role of N-acetyltransferase Phenotypes in Bladder Carcinogenesis: Pharmacogenetic Epidemiological Approach to Bladder Cancer." *Lancet* 2 (8303):842–45.

Doull J. 1980. "Factors Influencing Toxicology." In *Casarett and Doull's Toxicology* (2nd edition). New York: Macmillan Pubs., Co., Inc. pp. 70–83.

Ehrenberg, L., Moustacchi, E., Osterman-Golkar, S., Ekman, G. 1983. "Dosimetry of Genotoxic Agents and Dose/Response Relationship of Their Effects." *Mutation Research* 123:121–82.

Fukuto, T. R. 1990. "Mechanism of Action of Organophosphorus and Carbamate Insecticides." *Environmental Health Perspectives* 87:245–54.

Griffith, J., Duncan, R. C., Hulka, B. 1989. "Biochemical and Biological Markers: Implications for Epidemiologic Studies." *Archives of Environmental Health* 44:375–81.

Grouchy, J., and Turleau, C. 1984. *Clinical Atlas of Human Chromosomes.* New York: John Wiley and Sons. pp. 1–417.

Gupta, R. C., and Earley, K. 1988. "32P-Adduct Assay: Comparative Recoveries of Structurally Diverse DNA Adducts in the Various Enhancement Procedures." *Carcinogenesis* 9:1687–93.

Hirschorn, K., and Cohen, M. M. 1968. "Drug-Induced Chromosomal Aberrations." *New York Academy of Science* 151:977–87.

Jones, P. B., Galiazzi, D. R., Fisher, J. M., and Whitlock, J. P. 1985. "Control of Cytochrome P-450 Gene Expression by Dioxin." *Science* 227:1499–1502.

Leavell, B. S., and Thorup, O. A. 1976. *Fundamentals of Clinical Hematology 4 ed.* Philadelphia: W. B. Saunders Co.

Lemonnier, M. C., Julian, C., Ayme, S., Philip, N., Voeckel, M. A., Garnerre, M., and Mattei, J. F. 1990. "The Usefulness of the Level of Alpha-fetoprotein (AFP) and Electrophoresis of Amniotic Cetylcholinesterase for the Detection of Selected Congenital Malformations." *Journal of Gynecology, Obstetrics and Biological Reproduction* (Paris) 19(3):280–84.

Lerman, L. S. 1986. ed. *DNA Probes: Applications in Genetic and Infectious Disease and Cancer.* Cold Spring Harbor Laboratory. pp. 1–188.

Manz, A., Berger, J., Dwyer, J. H., Flesch-Janys, D., Nage, S., and Waltsgott, H. 1991. "Cancer Mortality Among Workers in Chemical Plant Contaminated With Dioxin." *Lancet* 338:959–64.

Maroni, M., Jarvisalo, J., and La Ferla, F. 1986. "The WHO-UANDP Epidemiologic Effects of Exposure to Organophosphorus Pesticides." *Toxicology Letters* 33:115–23.

McKinnon, P. J. 1987. "Ataxia-telangiectasia: An Inherited Disorder of Ionizing-Radiation Sensitivity in Man." *Human Genetics* 75:197–208.

Nemali, M. R., Reddy, M. K., Usuda, N., et al. 1989. "Differential Induction and Regulation of Peroxisomal Enzymes: Predictive Value of Peroxisome Proliferation in Identifying Certain Non Mutagen Carcinogens." *Toxicology and Applied Pharmacology* 97:72–87.

Okey,. B., Roberts, E. A., Harper, P. A., and Denison, M. S. 1986. "Induction of Drug-metabolizing Enzymes: Mechanisms and Consequences." *Clinical Biochemistry* 19:132–41.

Omenn, G. S., and Seboin, H. V. eds. 1984. "P-450 Systems." In: *Genetic Variability in Responses to Chemical Exposure: Banbury Report 16.* Cold Spring Harbor Laboratory. pp. 35–97.

Perera, F. P., Hemmminki, K., Young, T. L., Brenner, Y. D., Kelly, G., and Santella, R. M. 1988. "Detection of Polycyclic Romatic Hydrocarbon. DNA Adducts in White Blood Cells of Foundry Worker." *Cancer Research* 48:228–29.

Perera, F. P., Santella, R. M., Brenner, D., Poirier, M. C., Munshi, A. A., Fischman, H. K., Van Ryzin, J. 1987. "DNA Adducts, Protein Adducts and Sister Chromatid Exchange in Cigarette Smokers and Non-smokers." *JNCI* 79:449–456.

Reuterwall, C., and Albertini, R. J. 1990. eds. *Mutation and the Environment, Part C.* New York: Wiley-Liss, Inc. pp. 357–366.

Santella, R., Li, Y., Zhang, Y., Young, T., Stefanidis, M., Lu, X., Lee, B., Gomes, M., and Perera, F. P. 1990. "Immunologic Methods for the Detection of Polycyclic Romatic Hydrocarbon-DNA and Protein Adducts." In: *Genetic Toxicology of Complex Mixtures.* ed. M.D. Waters New York: Plenum Press. pp. 292–301.

Silk, D., King, J., and Whittaker, M. 1979. "Assay of Cholinesterase in Clinical Chemistry." *Annals of Clinical Biochemistry* 16:57–75.

Silverstein, R. M., and Bassier, G. C. 1968. *Spectrophotometric Identification of Organic Compounds.* New York: John Wiley & Sons, Inc. pp. 1–256.

Taplin, S. H., Thompson, R.S., Conrad, D.A. 1988. "Cost-Justification Analysis of Prenatal Maternal Serum-alpha-feto Protein Screening." *Medical Care* (Philadelphia). 26(12):1185–202.

Wilson, J. G. 1977. "Female Sex Hormones." In: *Handbook of Teratology, Vol. 1:* General Principles and Etiology. (eds) J.G. Wilson and F.C. Fraser. New York: Plenum Press. pp. 324-27.

World Health Organization. 1987. *WHO Laboratory Manual for the Examination of Human Semen and Semen-Cervical Mucus Interaction.* New York: Cambridge University Press, pp. 1–27.

8

Biomarkers in Environmental Epidemiology

Charles H. Nauman, Jack Griffith, Jerry N. Blancato, and Tim E. Aldrich

OBJECTIVES

This chapter will:

1. Discuss biomonitoring and its role in environmental epidemiology.
2. Describe markers of exposure and effect as they apply to environmental epidemiology.
3. Define Type 1 and Type 2 markers and their relationship to exposure and effect.
4. Review the application of biological markers to exposure assessment, public health, and methodological issues.

INTRODUCTION

As mentioned in Chapter 6, accurate exposure assessment is the **gold standard** for environmental epidemiology. Nonoccupational human exposures to toxic chemicals tend to be at low levels, from mixtures of suspect compounds found in consumer products, hazardous waste sites, agricultural pest control practices, combustion processes, and other sources. Even where exposures can be narrowed to one or a few chemicals, the low frequency of disease outcomes and the presence of confounding factors make an accurate assessment of exposure critical.

Biomarkers are becoming increasingly important in supporting the estimation of exposures for epidemiologic studies. Sensitive monitoring and laboratory analysis techniques now are being employed in what has been termed **biochemical** or **molecular epidemiology** to detect alterations in chemical structure or physiology that may qualitatively or quantitatively complement other measurements of exposure. These molecular events provide not only markers of exposure, but also may represent events in the continuum between exposure and disease. Associating these intermediate events mechanistically to exposure and to resulting disease is a research area of high current interest.

Biomarker (or biomonitoring) data are measures of an individual's internal exposure, or dose, and have several advantages over measures of exposure external to the body. Ambient or personal exposure measurement data provide information on the concentration of the toxic agent present in general or specific environments frequented by the individual. On the other hand, biomonitoring or biomarker data provide a confirmation and measure of the **bioavailability** of the contaminant of concern. Bioavailability relates to the completeness of absorption of a substance from environmental media. Biomonitoring data represent an integration of continuous and/or intermittent exposures coming from all sources, traveling through all environmental pathways, and passing into the body through all portals of entry (oral, inhalation, dermal).

Since biomonitoring data represent dose, they are more closely related to adverse health effects than are exposure monitoring data; thus, these data potentially are more valuable in the ultimate estimate of health effects, and in estimating the overall risks following single and multiple compound exposures. In certain instances these data can be used reconstructively to estimate previous or ongoing exposures.

Monitoring data on concentrations of compounds in environmental media (air, water, soil, food) provide estimates of general population exposure. Biological monitoring defines that portion of an exposure which has crossed the exchange boundaries of the body (skin, linings of the gut and lung). Tentatively, such biomonitoring measures can be used to identify excessive exposures before adverse health is manifest. Interindividual differences in accumulation of the substances of concern and in expression of biomarkers can be appreciable due to various host factors, such as variations in metabolic patterns. However, if groups (for example, by sex, age, health status) in a population are monitored, inferences can cautiously be made regarding the exposure of that population.

A variety of biomarkers are presently being evaluated to determine their usefulness as markers of exposure, internal dose, biologically effective dose, susceptibility to disease, and frank clinical disease. Understanding the basic mechanism of their formation, and the kinetics of their appearance and disappearance, is critical to their use and to being able to define the ultimate exposure-response relationships for environmental chemicals. Validation of biomarkers, to determine their specificity, sensitivity, and predictive power, is essential to providing meaningful measures of exposure and dose for use in environmental epidemiology.

DEFINITIONS, CLASSIFICATION OF MARKERS, AND CONCEPTS

Biomonitoring

Biomonitoring, or biological monitoring, has been described as the routine analysis of human tissues or excreta for direct or indirect evidence of exposure to chemical substances (Miller 1984), or physical agents. Biological monitoring, from its inception, has been a broad term that includes monitoring for any indication that exposure has occurred.

The organ and tissue sites that are potentially available for biomonitoring and for the study of biomarkers in humans is shown in Figure 8-1. Blood, the most readily accessible tissue for monitoring, also serves as a vehicle to distribute toxic agents throughout the body. Frequently, rather than monitor tissue in the organ of interest, a surrogate (blood or urine) is used, either because the monitoring process would be too invasive or too costly. For example, workers exposed to N-substituted aryl compounds (arylamines) such as benzidene and 2-aminonaphthalene are believed to be at increased risk of bladder cancer. If these workers also happen to be individuals with low levels of the genetically determined enzyme marker, N-**acetyltransferase (Nat)**, their risk of bladder cancer is enhanced. Clearly, we need a noninvasive

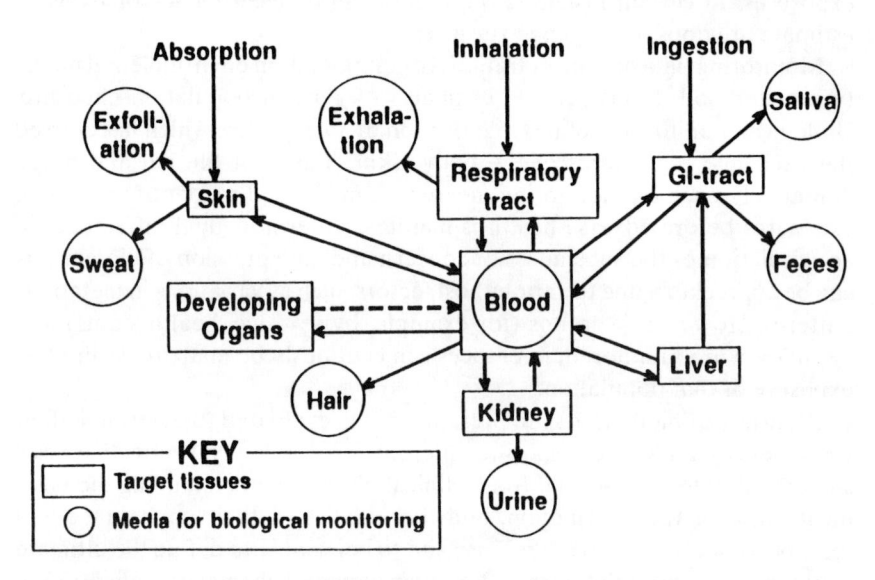

FIGURE 8-1. Sites of Uptake, Target Tissue, and Media for Monitoring Xenobiotic Compounds. (Adapted from Committee on Biological Markers, National Research Council. Biological "Markers in Environmental Health Research." *Environ Health Perspectives* 1987 77:3–9). Source: With permission, Hulka et al. 1990, Oxford University Press (New York).

marker if we are to determine the exposed workers susceptibility to disease by monitoring for this enzyme. There are laboratory assays for blood or urine available for Nat phenotyping.

The primary assumption, and an important one, is that the surrogate tissue is a true reflection of the organ site. This may or may not be the case, and must be judged in each instance on the basis of the type and duration of exposure, and the biological characteristics of the marker (for example, its persistence in tissue following exposure).

The National Research Council (NRC) has defined **biological markers as indicators signaling events in biologic systems or samples** (NRC 1987), and biologic markers as indicators of variation in cellular or biochemical components or processes, structures or functions which are measurable in a biological system or sample (NRC 1989a; NRC 1989b). The ideal biological marker would be pollutant specific (or generally responsive to toxicants in a complex mixture if that is the exposure of concern), available for analysis via relatively noninvasive procedures, detectable in trace concentrations, inexpensive to identify and measure, and quantitatively relatable to the degree of exposure in an exposure-response fashion. This ideal marker also would appear early in the exposure-to-health-effect continuum, have a totally nonadverse impact, and be quantitatively predictive of a health effect.

Biomarkers, biological markers, and biologic markers are, for all practical purposes, equivalent terms. However, a distinction currently is arising through common usage between these terms and the term biomonitoring. Biomonitoring now is tending to be equated to residue analysis (detection of parent compound or metabolite in tissue, fluid, or breath), while all other markers are included under the term biomarker. This is because residue analysis (a measure of body burden) is considered by some to be chemical dosimetry. For the purposes of this discussion, we define a biomarker as any measurable chemical, biochemical, physiological, cytological, morphological, or other biological parameter obtainable from human tissues, fluids, or expired gases, that is associated (directly or indirectly) with exposure to an environmental pollutant.

Biomarkers have been classified as those of exposure, susceptibility, and effect—and defined by the NRC Committee on Biologic Markers (NRC 1989a; NRC 1989b).

Biologic Marker of Exposure

A **biologic marker of exposure** is an exogenous substance or its metabolite or the product of an interaction between a xenobiotic agent and some target molecule or cell that is measured in a compartment within an organism. Those biomarkers of exposure having the most potential value are those nonadverse chemical, biochemical, or physiological changes that can be correlated with exposure to specific compounds.

Biologic Marker of Effect

A **biologic marker of effect** is a measurable biochemical, physiologic, or other alteration within an organism that, depending on magnitude, can be recognized as an established or potential health impairment or disease. Thus, biomarkers of effect are indicators of the functional capacity of the system, or an altered state of the system that can be measured.

Biologic Markers of Susceptibility

A **biologic marker of susceptibility** is an indicator of an inherent or acquired limitation of an organism's ability to respond to the challenge of exposure to a specific xenobiotic substance. Thus, biomarkers of susceptibility are indicators that the health of the system is especially sensitive to a xenobiotic. In epidemiologic studies, susceptibility biomarkers could improve the precision and strength of putative exposure–disease associations by avoiding the dilution effect that occurs in populations with a large proportion of nonsusceptible persons (Wilcosky and Griffith 1990).

Relationships to Dose and Effect

The interrelationships of these biomarkers can be evaluated effectively by reference to the **continuum of events** between exposure to a toxic compound and a resulting health effect, as indicated in Figure 8-2. The three intermediate events in this continuum include steps where absorbed (systemic) dose can be measured in tissues or fluids, effective dose can be measured or estimated in target tissues, and biological effects can be monitored via various tests and assays (see Chapter 6 for additional discussion on dose-response). The events that can be evaluated in the biological effect step of the continuum are numerous and varied; these biological effects have been categorized as those of early response, and those of altered structure/function (NRC 1987).

Exposure biomarkers are represented in Figure 8-2 as being associated most closely with absorbed dose and target dose, while effect markers are relegated to the final events in the continuum. In reality, however, the distinction between these two types of biomarkers is not so clearly defined. Many markers of effect are also markers of exposure; however, we know enough about only a few markers of exposure that they can serve as markers of health effect. Examples of this latter type include carboxyhemoglobin and

```
                    ┌─────────────────┐
                    │  Susceptibility │
                    │    Biomarkers   │
                    └─────────────────┘
          (a)         (b)    (c)         (d)
```

| Applied Dose (Exposure) | Absorbed Dose | Delivered or Target Dose | Biological Effect | Health Effect |

| Ambient/Personal Measurements | Exposure Biomarkers | Effect Biomarkers |

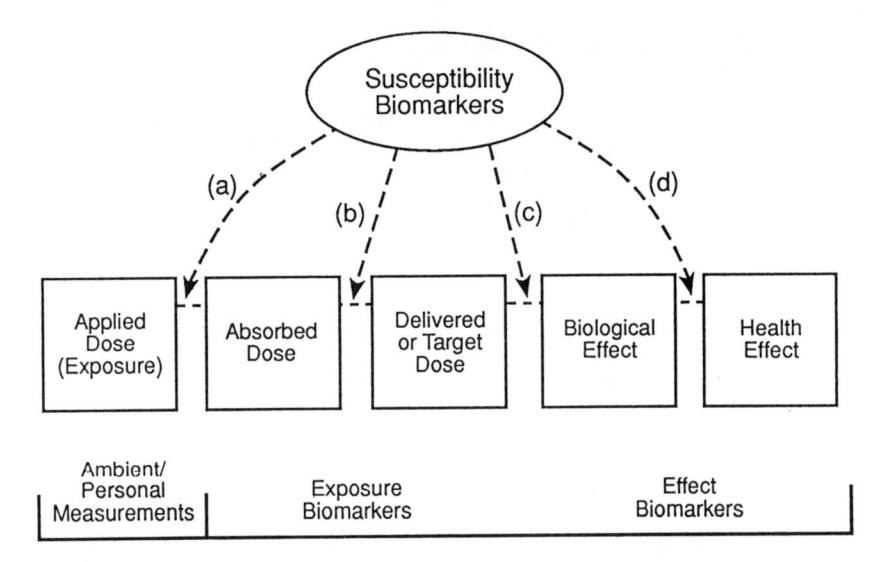

FIGURE 8-2. Relationships Between Biomarkers of Exposure, Effect, and Susceptibility, and Events in the Continuum From Exposure to Health Effect.

blood lead levels; these can be equated to known impairment or health outcomes. Susceptibility biomarkers are genetic predispositions, or acquired conditions brought on by disease or occupational/environmental exposure, which can affect the progression at each internode (segment) across the continuum. Susceptibility markers are considered to be effect modifiers in epidemiologic terms (Hulka 1990).

There are many potential susceptibility markers at each internode. The following are generic examples. At internode (a) any alteration in cellular membrane permeability could enhance passage of a xenobiotic and alter the absorbed dose (and ultimately enhance a health impact). Induction of carrier proteins at internode (b) could enhance dose to sensitive tissue or cellular targets. A decreased level of repair enzymes operating at internode (c) might modulate the biological effect, thus altering the severity of the health effect. And at internode (d) a (for example, immune) system dysfunction could accelerate progression to a disease state.

SURVEY OF BIOMARKERS

Various schemes can be used to categorize the various types of biomarkers that have been described. One such representation is provided in Table 8-1.

158 Environmental Epidemiology

TABLE 8-1. Categorization of Biomarkers.

Type 1. Indicators of body burden.
  A. Residues—Parent compound or metabolite in tissue or fluid or breath.
  B. Interactive Products—Adducts (nucleic acid and protein), receptor-xenobiotic complexes, conjugate complexes.

Type 2. Response to absorbed dose.
  A. Physiologic response.
    — Isozyme induction.
    — Enzyme inhibition.
  B. Genotoxic response.
    — Sister chromatid exchange.
    — Chromosome aberration.
    — Micronucleus.
    — DNA strand breakage.
    — Mutation.
  C. Organ level response.
    — Organ dysfunction.
    — Hyperplasia.
    — Polyps.
  D. Oncogenic response.
    — Oncogene activation.
    — Oncogene products.
    — Tumor.

Type 3. Predispositions/acquired conditions.
  A. Aryl hydrocarbon hydroxylase.
  B. Acetylator type.
  C. Glutathione S transferase.

TYPE 1 BIOMARKERS

Type 1 biomarkers often include both markers of exposure and markers of effect. As previously indicated, residue analysis (or pure chemical dosimetry) is a measure of body burden, indicating that exposure has taken place. Rarely does this type of measurement reveal the total amount of the analyte (substance or material analyzed) in the entire body, and it is most often given in units of mass of analyte per unit mass of tissue or volume of fluid. For more than a decade epidemiologists have attempted to estimate human exposure by making measurements of residues in tissues and fluids of exposed persons. For example, individual exposure to polychlorinated biphenyls (PCBs) has been monitored by testing for PCB residues in human tissue and fluids (Wickizer et al. 1981; Brilliant et al. 1978). Over many years, blood lead has been used extensively as an exposure marker for lead pollution (McLaughlin et al. 1983; Grobler et al. 1986), and there have been successful attempts to associate blood lead residues with hypertension (Harlan et al. 1985) and the risk of cardiovascular disease (Pirkle et al. 1985).

Interactive products have been described as markers of both exposure and effect. The interactions that occur to form these markers are biological effects; however, not all biological effects are adverse, and if a relationship cannot be established with adverse health impact a biological effect marker is most useful as a marker of exposure. DNA adducts perhaps come closest to being indicators of an adverse health impact because some can be associated with specific mutations that are thought to lead to cancer (Loveless 1969). The applicability of protein and DNA adducts will be discussed in more detail in a later section. In general, Type 1 biomarkers tend to be compound specific, in contrast to many of the Type 2 markers.

Type 2 Biomarkers

Type 2 biomarkers all represent effects, but, as noted, many of these also serve as markers of exposure. Essentially all Type 2 markers are chemically nonspecific, capable of being expressed as a result of exposure to a variety of chemical substances. Thus, these markers may be most valuable for ascertaining exposure to mixtures of compounds.

Physiological Response Markers

Numerous physiological responses can be used as indices of exposure. A commonly used example is that of inducible enzymes of the cytochrome P_{450} mixed-function oxidase (MFO) system that mediates Phase I metabolic reactions of xenobiotics. This is a general response that may already be elevated due to previous or concurrent exposures. Examples of this will be given in a later section.

Following exposure to organophosphate (OP) and thiocarbamate pesticides red blood cell (RBChE) or plasma cholinesterase (PChE) depression may serve as an index of exposure. Griffith and Duncan (1985) measured PChE among 73 spray-season citrus workers and 55 citrus harvesters. The observed PChE mean values were found to be below the values, at statistically significant levels, found in unexposed controls. A review of cholinesterase activity among workers occupationally exposed to OP pesticides has facilitated the development of criteria for protecting workers prior to onset of neurological symptoms (Coye et al. 1986). Bhatnagar et al. (1982) compared PChE values in 75 pesticide factory workers with 15 nonexposed controls. The statistically significant PChE values of the workers were reduced compared to controls, and reflected heavy OP exposure. Numerous other examples exist of the use of cholinesterase depression as a biomarker of exposure (for example, see Wicker et al. 1979).

Genotoxic Response Markers

There are a multitude of genetox assays that may prove useful as biomarkers. A few generic assays will be discussed here.

Sister Chromatid Exchange. Perry and Evans (1975) demonstrated that sister chromatid exchanges (SCEs), originally identified by Taylor (1958), could be induced *in vitro* by chemical carcinogens. The use of this technique has been extended to the investigation of mutagenic potential of carcinogens in humans. Sister chromatid exchanges have frequently been used as a biomarker for assessing the mutagenic potential of chemicals (Allen et al. 1983; Stolley et al. 1984). Sister chromatid exchanges occur during cell replication as a chromosome duplicates its genetic material forming sister chromatids (a pair of chromosomes) attached at the centromere. Sister chromatids, through DNA breakage and rejoining, can exchange apparently identical DNA segments without altering the viability or functioning of a cell (Wilcosky and Rynard 1990). Since SCE measurements are affected by chemical exposures, it is critical that careful occupational, home, and medical histories be developed to characterize exposure in order that potential confounding can be addressed. Although SCEs are not known to result in a disease endpoint, they are a sign that chromosomal alteration has occurred.

Chromosomal Aberrations. Early in the 20th Century, Boveri (1914) proposed the somatic mutation theory of cancer, and implicated chromosomal aberrations by suggesting that a "wrongly combined chromosomal complex" in a somatic cell was the heritable cause of abnormal proliferation (Schwartz 1990). Chromosomal abnormalities may be classified as structural or numeric. Structural damage can be identified at cell division during metaphase (chromosome-type and chromatid-type). Chromosome-type damage occurs during exposure to ionizing radiation or mutagenic chemicals during the G_0 or G_1 phase of mitosis. Chromatid-type damage occurs when exposure to a mutagen takes place during the S or G_2 cycle. Chromatid-type aberrations result from errors in DNA replication during the S phase of mitosis, and thus are not primary genetic lesions. There are seven chromosome-type aberrations: terminal deletions, minutes, acentric rings, centric rings, inversions, reciprocal translocations, and dicentric or polycentric aberrations; and nine classes of chromatid-type aberrations: gaps, breaks, minutes, acentric rings, centric rings, inversions, symmetrical exchanges, isochromatid aberrations, and asymmetrical interchanges. Software recently has been developed for the consistent analysis of chromosome aberration data (U.S. EPA 1990a).

The genotoxicity of environmental or occupational exposures has been assessed through the analysis of chromosomal aberrations in numerous studies. Recently, Garry et al. (1989) found that applicators exposed to phosphine, a common grain fumigant—or to phosphine and other pesticides—had significantly increased stable chromosome rearrangements, primarily translocations in G-banded lymphocytes, when compared with unexposed controls. In the example shown (Figure 8-3), a section of chromosome 8 has broken off and joined chromosome 10 (see arrows). This translocation occurred at breakpoint q12 and q23 on the chromosome. Less stable alterations, including chromatid deletions and gaps, were significantly increased during, but not after, the application season.

Micronuclei. The micronucleus assay detects in interphase cells the presence of chromosome fragments that have been excluded from the nucleus. These fragments are the result of chromosome breakage that occurred in the previous mitotic cycle, or can be entire chromosomes that have lagged during the previous cell division due to an impact on the mitotic spindle, representing aneuploidy. The *in vivo* micronucleus assay has been modified and enhanced in recent years and represents a rather standard procedure (Heddle et al. 1983; Jauhar et al. 1988). Software also is available for the analysis of micronucleus data (U.S EPA 1990b).

DNA Strand Breakage. A DNA migration assay, the Single Cell Gel (SCG) technique, has been developed in recent years as a sensitive measure of single strand breaks (Singh et al. 1988). This assay makes use of the migration of DNA strands after unwinding in an electrophoretic field as a measure of genetic damage resulting from exposure to genetically active chemicals. To find the fragment with the sequence of interest, one first denatures the DNA fragments on the gel with alkali, which makes it possible to render the normally double stranded fragments single stranded so that the fragments can hybridize with complementary fragments of DNA. When DNA is hybridized, two single-stranded DNA molecules, or a single-stranded DNA molecule and an RNA molecule join to form a double stranded molecule (Schwartz 1990).

The degree of migration is proportional to the amount of damage to the DNA. Advantages of the technique include its sensitivity, the requirement for a very small sample size (a finger prick provides a sufficient blood sample), and the fact that intercellular variation in response can be monitored. The SCG assay has been refined and evaluated with a variety of cell types from a variety of mammalian species, including humans (Singh et al. 1990). Peripheral blood lymphocytes have been the cells most frequently monitored; however, liver, brain, and bone marrow have been successfully evaluated in rodent species serving as surrogates for human exposures near a hazardous waste site (Tice 1990).

Mutation. A mutation is a change in the nucleotide sequence of DNA, or it may be defined more generally as a permanent change in genotype other than one brought about by genetic recombination (NRC 1983). These

LABORATORY OF ENVIRONMENTAL MEDICINE AND PATHOLOGY
421 SOUTHEAST 29TH AVENUE
MINNEAPOLIS, MINNESOTA 55414

SUBJECT Du 174

DATE 12-8-87   TYPE OF BANDING G

FILM NO. 6-5   SLIDE/LOCATION B1-3 (153.2 X 24.1)

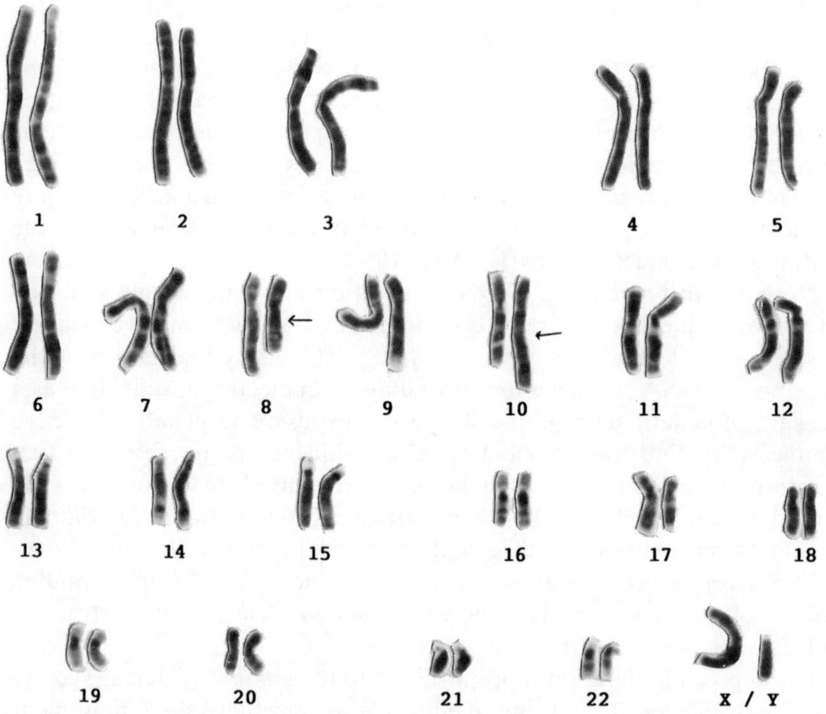

ANALYSIS 46,XY,t(8;10)(q12;q23)

FIGURE 8-3. Genetic Translocations on Workers Exposed to the Fumigant Phosphine.

changes represent point mutations (base pair substitutions or frameshifts), small deletions, or amplifications. A mutation is heritable in the sense of being replicated through somatic cell divisions, or it can be passed to offspring if the mutation is present in germ cells.

A large battery of mutation assay systems exists (Brusick 1980). These range from the Salmonella (Ames) assay, to fungal assays (e.g., Nuerospora and Aspergillus), assays in plant systems such as Tradescantia and Zea, classic assays in Drosophila, mammalian cell culture assays (for example, the mouse lymphoma L5178Y cell assay, the human HeLa and other cell assays), as well as mammalian *in vivo* assays such as the mouse specific locus test.

DNA/Protein Adducts. Adducts are relatively stable complexes of reactive chemicals and cellular macromolecules that contain one or more covalent bonds between the moieties. Zarbl et al. (1985) and Bishop (1987) have demonstrated a mechanism for linking adducts, mutations, and carcinogenesis. Although the referenced work suggests that DNA adducts may play a role in the development of cancer, epidemiologic investigation of these associations has been limited due to the lack of sensitive and specific assays that could be applied to a community population. However, new methods present considerable promise in the investigation of DNA adducts. For example, the sensitivity of ^{32}p-postlabeling is such that it can detect a single adduct from an aromatic hydrocarbon per cell, or as little as one microgram (μg) of adduct per 100 kg of tissue. Assays can be performed with as little as 10 μg of DNA, which is the product of about 10 mg of tissue.

The relatively easy availability of lymphocytes makes them very attractive cell populations for the analysis of DNA adducts as well as for cytogenetic analysis. Since the life span of a lymphocyte may vary from only a few days to as much as 20 years (Leavell and Thorup 1976), they may be useful for estimating the effects of long-term chemical exposures, although the rate of DNA repair becomes increasingly important in determining adduct levels in longer lived cells. As Perera et al. (1987) have suggested, since lymphocytes are in contact with many body tissues, it is conceivable that they may provide an integrated measurement of exposure.

Erythrocytes (red blood cells) have no nucleus and thus, no chromosomes. Analysis of adducts in these cells is limited to protein adducts of hemoglobin. However, since hemoglobin is relatively easy to obtain, and since the life span of an erythrocyte is long (120 days) relative to other cells, it is possible to investigate cumulative dosage in chronic exposure studies using red blood cells (RBCs). Although hemoglobin adducts are not believed to play a direct role in tumor induction, they may effectively serve as surrogates for other processes. For example, in humans, small alkylating agents such as ethylene oxide are uniformly dispersed throughout the body, and the amount of hemoglobin alkylation generally approximates the dose to DNA (Ehrenberg et al. 1983). In certain instances we know that the formation of protein adducts is proportional to DNA adducts; this is true especially in cases where the ultimate electrophile is the same for both types of adduct (Schnell and Chiang 1990). However, this is not necessarily the case for all adduct–compound combinations, and the relationship between hemoglobin and DNA adducts must be determined experimentally for each particular compound.

Ideally, adducts should be measured in target tissues or cells. Sampling target tissues in many instances requires very invasive procedures, not practical for epidemiologic studies. Thus, blood cells or plasma (for albumin adducts) often are chosen even though these may be remote from the tissue or cell where we wish to measure the effective (target) dose. The question of how to relate these surrogate measurements to adduct levels in target tissues is difficult. However, help with this question is potentially available through the development of pharmacokinetic and/or pharmacodynamic modeling. Through this process, consideration of factors such as transport to various tissues, intertissue differences in concentrations of metabolizing enzymes, and the presence of repair enzymes induced by previous exposures may increase confidence in estimates of adduct levels in target tissues.

Since the quantity of adducts present in tissue may integrate exposures occurring over time, and reflect intraindividual differences in the pharmacokinetics and metabolic activation of agents capable of interacting with DNA and protein, the detection and quantification of these adducts should improve dose-response relationships and provide an important means to quantify the biologically effective dose of environmental carcinogens. Also, by quantifying adducts at the molecular level following measurable dosage in different species, adducts can help researchers understand the comparability of doses received by humans and experimental animals.

Although methods in laboratory analysis have improved the identification of DNA adducts in the past few years (Henderson et al. 1989), most, including ^{32}p-postlabeling, are still in experimental stages of development. Other methods for detecting carcinogen-DNA adducts are available, e.g., the enzyme-linked immunosorbent assays (ELISA) and ultrasensitive enzymatic radioimmunassays (USERIA) that are perhaps more effective at quantifying exposure than ^{32}p, but are considered to be less sensitive (Wolff 1985). The striking advantage in using immunoassays to measure exposure to carcinogen-producing chemicals is the ability to identify adducts at extremely minute levels in tissue samples so small as to contain only 1 or 2 µg quantities of DNA (Haseltine et al. 1983).

A potential problem in using DNA adduct detection techniques is that cells also have DNA repair capacity that may remove most of the exposure-related adducts before an effective count can be taken, thus making it difficult to monitor populations for chronic low-level exposure (Wolff 1985). There is also concern that these cellular biomarkers are not predictive of future disease.

Organ-Level Response Markers

Organ-level responses, or organ dysfunctions, provide biomarkers of exposure and effect. These responses often are equated to noncarcinogenic effects of environmental pollutants, being related to health effects, and indicate nonspecific chemical exposures. Diverse chemical exposures can lead to the same or similar overall organ responses. Although these types of markers can indicate exposure and potential risk of disease, they cannot be used as quantitative markers to identify the specific chemicals responsible. Work is progressing to develop organ-system biomarkers with more specificity.

In the context of environmental epidemiology, a biomarker of organ damage or dysfunction is a measurable alteration in cells or cell products caused by an abnormal effect on a particular organ system (CDC/ATSDR 1990). The organ systems of most current concern include pulmonary, reproductive/developmental, neurologic, immunologic, renal, hepatic, and integumentary. Organ-level responses can range from reversible, transient changes with no health impacts to permanent changes that can be associated with morbidity and/or mortality. Preclinical biomarkers are associated with the early end of this range, and obviously are of the most value for intervention. Thus, organ-level biomarkers tend to be tests that permit a general assessment of tissue health, including early steps of toxicity to the organ.

The list of tests or assays potentially yielding useful biomarkers for organ-level systems is quite long. The tests have been described in a number of recent documents, including a listing of sentinel health events (Rothwell et al. 1991); biomarker panels for renal, hepatobiliary, and immune systems (CDC/ATSDR 1990), and the NRC publications (NRC 1989a; NRC 1989b) addressing pulmonary and reproductive toxicity biomarkers (a companion document on biomarkers for neurotoxicology is nearing completion). Individual biomarkers for these systems will not be discussed here.

One organ-level system for which biomarkers are not often addressed is the largest organ of the body, the integument (see Chapter 6 for a discussion on exposure and absorption through the skin). As with the other organ systems, numerous indicators of the health of this system are available. Dermatitis is one general response that can be very indicative of environmental exposure. The symptoms (which serve as biomarkers) of dermatitis are erythema, vesicles, oozing, and crusting. As the dermatitis becomes chronic, scaling and thickening of the skin occurs. As the disease progresses, microscopic and macroscopic vesicles (small blisters) and bullae (large vesicles) are formed.

Contact dermatitis is a very common occupational health problem. If lesions are confined to exposed areas of the body (for example, hands and face), there is a strong likelihood that contact with an environmental agent is responsible. If the lesions are distributed evenly on the exposed surfaces, the possibility of exposure to an airborne agent should be considered. If, on the other hand, the lesions are patchy or linear, consider chemical exposure, or perhaps exposure to a paint as the causative agent. Allergic mechanisms also cause contact dermatitis. Chemicals become antigens after joining with skin proteins. After antibodies have been formed to the protein-chemical conjugate, an eczematous reaction will occur on subsequent exposures to the chemical, thus providing an efficient biomarker of concurrent exposure.

Oncogenic Response Markers

One special class of response markers includes oncogene activation and oncogene protein products. The proto-oncogenes that are normally present in all cells are critical to normal cellular growth and development. Proto-oncogenes can become (inappropriately) activated, and yield oncogenes. Through processes that include the production of abnormal or excessive normal products, or the production of an ill-timed normal product, these oncogenes direct the malignant transformation of their host cells. Activated oncogenes occur in mammalian tumors of all types and in premalignant lesions, suggesting that oncogenes belong to a step in the pathway to cancer that is sufficiently early to permit clinical intervention (Perera et al. 1991; Brandt-Rauf 1991).

Proto-oncogenes can be activated by a number of processes including amplification, mutation, and chromosomal translocations. In one study the number of copies of the N-myc oncogene was found to be related to neuroblastoma tumor progression in all tumor stages (Seeger et al. 1985). Thus, identification of an activated (amplified) oncogene in an early tumor stage (where the tumor is confined to the organ or structure of origin, with no metastases) provides an opportunity for meaningful intervention.

Oncogene products also can be valuable indicators of incipient oncogenicity. Stock et al. (1987) immunochemically isolated a p55 protein from the urine of bladder cancer patients with transitional cell carcinoma, and compared its concentration with that for noncancer controls. Levels of expression of the p55 marker tended to correlate with tumor stage and grade. Disease progression at the time of expression was late when this marker was manifest, making its value in intervention minimal. However, the fact that this technique may be performed with a product in urine illustrates that all oncogene studies do not require invasively obtained tumor tissue. Also, the fact that none of the control patients showed elevated levels of the p55 protein illustrates that the false positive rate is low, and thus the technique potentially could be used as a screen for the presence of oncogene activity in urine.

Type 3 Biomarkers

Susceptibility is a particularly important factor to be aware of in environmental epidemiology (Omenn 1982). Relationships between exposure and biological effect or disease can be modified in that portion of a population that is susceptible, and the association between exposure and disease for the susceptible subpopulation can be masked or weakened in any study that does not consider interindividual susceptibilities. Susceptibility biomarkers should be used in stratifying susceptible subpopulations so that an estimate of disease risk for that group (and for the nonsusceptible group) can be made more accurately.

Type 3 biomarkers include all markers that may indicate the susceptibility of the host to enhanced impact following certain chemical exposures. Susceptibility biomarkers are independent of the exposure being evaluated, and are of two types: acquired and genetically determined (Vine and McFarland 1990). Acquired susceptibility indicators include age, diet (for example, vitamin intake, alcohol consumption), lifestyle, previous exposure to toxic compounds, and previous infections and diseases. Genetically determined biomarkers of importance to environmental epidemiology are numerous; three examples will be discussed here.

The first of these is the inducible enzyme aryl hydrocarbon hydroxylase (AHH). Aryl hydrocarbon hydroxylase is the form of cytochrome P_{450} that is induced by compounds such as 3-methylcholanthrene or benzo(a)pyrene (BAP), and is especially active in enhancing the biotransformation (activation) of polycyclic aromatic hydrocarbons (PAHs) in Phase I (early stage) metabolism. Genetically controlled differences in AHH activity are thought to account for the differences in metabolite binding levels to DNA between individuals who smoke similar amounts of tobacco (Harris et al. 1978). Thus, relatively lower activity levels of AHH may confer a lower susceptibility to those individuals exposed to PAHs.

A second example is N-acetyltransferase, a noninducible enzyme that functions in detoxifying aromatic amines such as benzidine and 4-aminobiphenyl by acetylation. Acetylation of arylamines occurs in two stages:

1. An acetyl group, CH_3CO, from acetyl coenzyme (cofactor) A is transferred to NAT, producing an acetyl-NAT intermediate (acetyl-N-acetyltransferase); and
2. The intermediate combinds with the arylamine, yielding an acetylated arylamine and regenerating NAT (Vine and McFarland 1990).

Two phenotypes are manifest in human populations, one with low acetyltransferase activity (slow acetylators), and one with high enzyme activity (fast acetylators). Those individuals who are slow acetylators are at increased risk of bladder cancer because they can not detoxify aromatic amines as quickly as fast acetylators (Lower 1982). The ratio of slow to fast acetylators varies by ethnic group, with Caucasians and blacks of African origin having approximately equal numbers of slow and fast acetylators.

A third example that functions in Phase II (late stage) metabolism to detoxify xenobiotics is the enzyme glutathione-S-transferase (GST). This family of isoenzymes catalyze the conjugation of reactive intermediates such as epoxides formed during Phase I (early stage) metabolism, with glutathione scavengers occurring naturally in the body. About 50 percent of Caucasians genetically are deficient in one of the isozymes of this family. Those individuals with a low capacity for producing the enzyme have a reduced capacity for conjugating electrophiles, and are more at risk for a mutational event that could be associated with cancer. The increased frequency of this GST deficiency observed in lung cancer patients indicates that this deficiency is a genetic marker of cancer susceptibility in humans (Seidegard et al. 1986). In addition, this deficiency has been shown to be a marker of susceptibility to increased chromosome damage by certain agents (Wiencke et al. 1990). The potential of GST, as well as that of AHH, as markers of susceptibility are being further evaluated (Perera et al. 1991).

APPLICATION OF BIOMARKERS TO EXPOSURE ESTIMATION

Several investigators have suggested that the failure to accurately assess exposure has reduced the effectiveness of epidemiologic research (Clarkson et al. 1983; Heath 1983). In fact, it has been suggested (Perera and Weinstein 1982) that the failure to accurately quantify exposure is the **Achilles heel** of traditional epidemiology, and that poor exposure assessment promotes misclassification among exposed and comparison cohorts. Recent research to develop and apply exposure biomarkers has provided good scientific evidence to suggest that biological monitoring tools will become increasingly useful in environmental epidemiology. The effectiveness of biomarkers in exposure assessment potentially can be enhanced by employing a suite or battery of appropriate biomarkers. Figure 8-4 presents the potential relationships (levels of occurrence and persistence in the body) between various biomarkers after a single exposure to a biologically active compound, as adapted from Henderson et al. (1989).

This illustration is instructive from several points of view (Figure 8-4). If one is faced with selecting a biomarker for use in an epidemiology study and exposure duration is expected to be short, these relationships point out the time frame after exposure in which collection of a particular sample would yield useful biomarker data. For example, urinary metabolites would be most useful if the samples were collected within days of exposure. Albumin adducts should be analyzed in blood drawn within two to three weeks of exposure, since we know the half-life of albumin in the body is about 20 days. Similarly, hemoglobin adducts should be present in RBCs over their 120-day lifetime. The timing for cellular and urinary DNA adducts will be compound and adduct specific, dependent on the rate of repair and turnover of the host cell population.

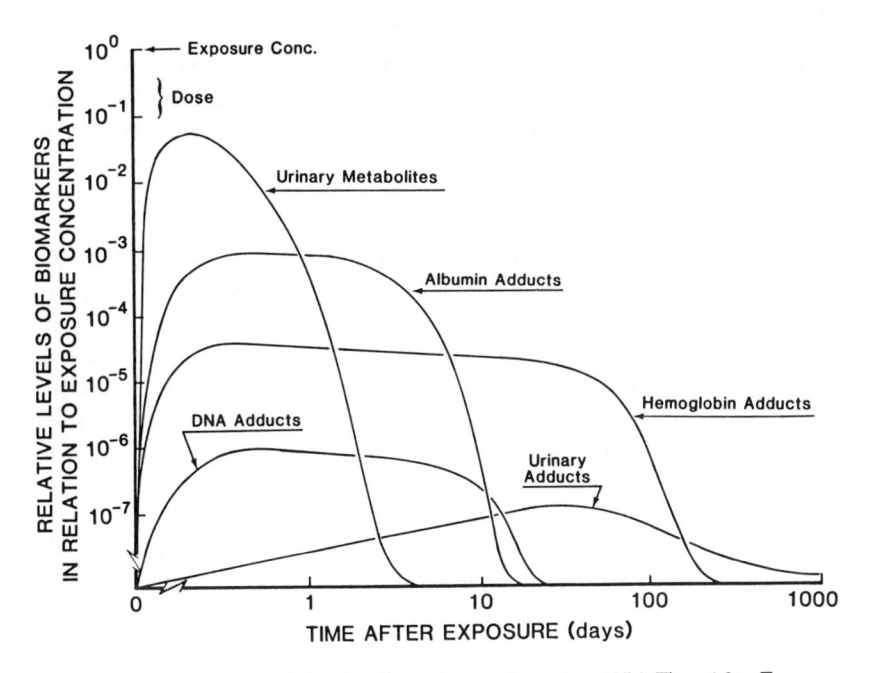

FIGURE 8-4. Potential Relative Levels of Various Biomarkers With Time After Exposure. Source: Adapted from Henderson et al. 1989.

The different levels of the various biomarkers in Figure 8-4 also suggest that if a suite or battery of biomarkers such as those represented were selected for use, the relative levels could be used to estimate when exposure occurred. For example, if all five markers (shown in Figure 8-4) were assayed in samples taken at less than one time unit after exposure, the resulting profile would include high relative levels of urinary metabolites, thus suggesting very recent exposure. At about five time units after exposure, the profile of biomarkers would include albumin, hemoglobin and DNA adducts in that rank order—indicating that exposure had occurred more than a short time period ago. Only hemoglobin, urinary protein, and DNA adducts would remain at 50 to 100 time units after exposure.

Exposures encountered in the environment, or occupationally, often are intermittent or continuous. For a continuous exposure regimen the curves in Figure 8-4 could peak and remain flat, representing a steady state. In this case, a group of biomarkers would assure detection of the exposure and would provide an instructive profile of marker levels that potentially could allow modeling of the level of exposure and the body burden of particular compounds.

APPLICATION TO PUBLIC HEALTH

As noted earlier, the ultimate objective of research in the area of biomarkers is to develop markers that represent stages early in the continuum that can be related (preferably quantitatively) to clinical outcome. (Another important objective can be to reveal that exposures are taking place, so that the source[s] of exposure may be eliminated or curtailed). This includes not only markers of exposure and/or effect, but those susceptibility markers that may give clues to the severity of health impacts that may follow exposure to certain compounds. Thus, the overall objective of using biomarkers is to avoid unnecessary disease, disability, and premature death.

Albertini (1990) has illustrated this objective in Figure 8-5. Early indicators of potential problems permit us the opportunity to invoke preventive measures to forestall any future health impact (that is, intervention that might include treatment to correct reversible damage and/or exclusion from future exposure). As represented here, hazard indicators could include the Salmonella assay, the L5178Y lymphoma assay, evaluation of mutational spectra in microorganisms, and rodent bioassays for suspect chemical compounds; all of these would alert investigators to potential problems for exposed individuals.

The exposure markers indicated in Figure 8-5 might include adducts to nucleic acid or protein, receptor-xenobiotic complexes, and conjugates appearing in urine. These markers are developed through animal studies and then applied in exposed human populations. Early effect markers might

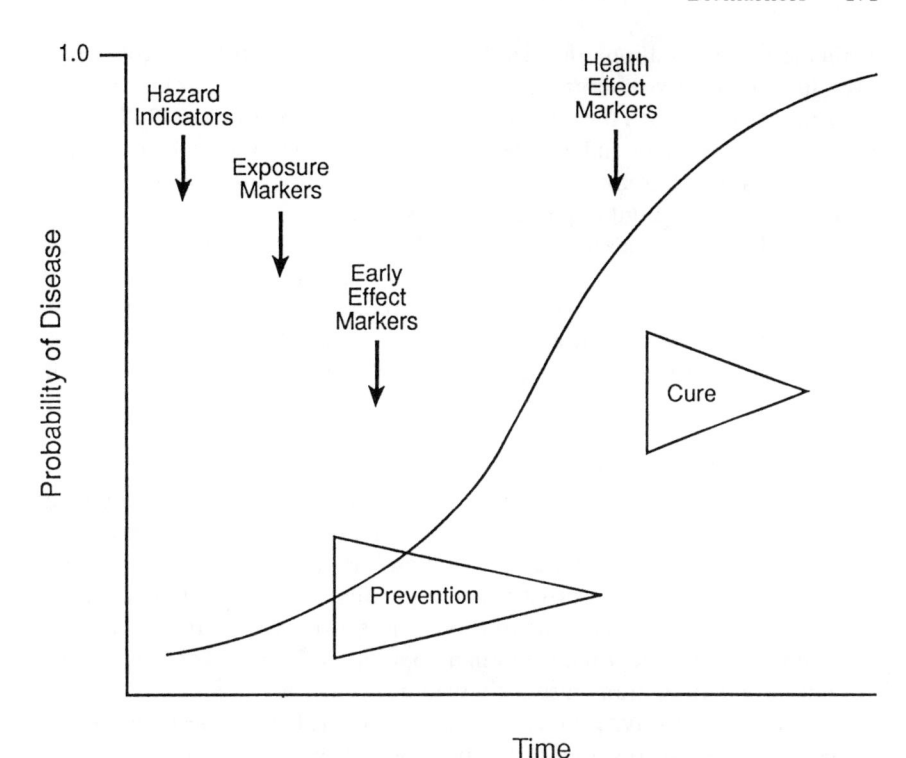

FIGURE 8-5. Time Course of the Effectiveness of Various Indicators of Hazard, Exposure, and Effect in Disease Intervention. Source: Adapted from Albertini 1990.

include any of several *in vivo* genotoxicity assays, or the expression of gene products—perhaps representing oncogene activation—in animals and humans. Health effect biomarkers that present themselves later in the continuum require intervention in the form of attempting a cure. Examples of markers at this point on the probability curve include tissue hyperplasia, polyps, and tumors. Potentially, the curve representing the probability of disease can be lowered at later time periods if early hazard indicators and biomarkers of exposure and effect are successfully developed and used.

METHODOLOGICAL CONSIDERATIONS

Epidemiologically, a biomarker may serve as the dependent variable in one design application and as the independent variable in another. For example, in a cross-sectional study designed to observe the effects of phosphine (a chemical fumigant) exposure on the rate of SCE in workers employed in grain elevators, the biomarker could serve as the dependent variable. However, in a prospective cohort study designed to investigate the relationship of phosphine exposure to the occurrence of cancer, the biomarker (SCE) could effectively serve as the independent variable. Thus, a biomarker can serve to mark exposure, thereby providing the potential for intervention, or as a predictor of subsequent impairment or disease.

Viewed as a dependent variable, the biomarker could serve to promote protective measures, up to and including modification or termination of hazardous exposure. When the biomarker is considered to be an independent variable, it may be viewed as the predictor of an effect and may promote understanding of the relative contribution of its precursor exposures to the etiology of the disease or effect in question.

Whether a biomarker is to be used as a dependent or independent variable, it should be judged in terms of its validity (that is, sensitivity and specificity) and its reliability (that is, the reproducibility of the reported results if the study were repeated elsewhere, under similar circumstances). In attempting to use biomarkers in epidemiologic studies, it is important to clearly understand the meaning of "sensitivity" and "specificity" as they relate to laboratory methods, and the meaning of those terms as they refer to the ability of the biomarker to detect the event in a population (Table 8-2; see Chapter 3 for additional discussion on sensitivity and specificity).

Laboratory sensitivity to detect a biomarker refers to the ability of the detection system to respond in the presence of the biomarker; it is usually ascertained by testing serial dilutions of the biomarker. If the assay can detect a biomarker at levels that are thought to be below an effect threshold with a small false negative rate, it is generally said to be **highly sensitive.** Laboratory specificity refers to the ability of the detection system to fail to respond when present with the assay medium only (for example, in the total absence of the biomarker). When the false negative rate of the assay is small, it is generally said to be **highly specific.**

TABLE 8-2. Population Values of Biomarker Sensitivity and Specificity for Detecting or Predicting an Event.*

	Event		
Biomarker	Present (E+)	Absent (E−)	Total
Present (M+)	bp	$(1-a)(1-p)$	m
Absent (M−)	$(1-b)p$	$a(1-p)$	$(1-m)$
Total	p	$1-p$	1

Notes: b = sensitivity, a = specificity, m = biomarker frequency, p = event frequency.
* Adapted from Khoury et al.

It is useful, when trying to detect or predict the event in a population, to arrange the distribution of subjects in a fourfold table according to the presence or absence of the biomarker and the presence or absence of the event. Sensitivity is then estimated as the ratio of the number of subjects positive for both the biomarker and the event to the number of subjects with the event. Specificity is estimated as the ratio of subjects negative for both the biomarker and the event to the subjects negative for the event. Table 8-2, adapted from Khoury et al. (1985), shows the relationships between population values of sensitivity, specificity, disease frequency, and biomarker frequency. The parameters shown in Table 8-2 arise from a multinomial distribution that governs the probability that a particular subject will be classified into one of the four cells of the table. Thus, there are three degrees of freedom associated with Table 8-2: If three of the cells are specified, then the other is algebraically determined; also, if three of the four parameters (a, b, m, p) are specified, then the fourth is determined algebraically. As shown by Khoury et al., the four parameters are related by

$$(a-1)(1-p) - bp + m = 0. \qquad [1]$$

The algebraic relationship is important in that it indicates a relationship (mediated by biomarker frequency and event frequency) between sensitivity and specificity of a biomarker for disease or exposure in the population in contrast to the lack of a relationship between laboratory sensitivity and specificity for detecting the biomarker. The defining equations for biomarker sensitivity and specificity, derived from equation [1], are

$$b = [(1-p)/p]a + (p + m - 1)/p \qquad [2]$$
$$\max(0, m + p - 1)/p \leq b \leq \min(m/p, 1)$$

$$a = [p/(1-p)]b - (p + m - 1)/(1-p) \qquad [3]$$
$$\max[0, 1 - m/(1-p)] \leq a \leq \min[(1-m)/(1-p), 1]$$

Although a laboratory method to detect a biomarker can be used in any population, either to aid in improving exposure classification or to study the association of the biomarker and the event, the sensitivity and specificity of the biomarker to detect the event are not necessarily generalizable from one population to another. This is best illustrated with data from the literature. Milby et al. (1964) published data on the relationship of depressed ChE and symptoms of OP insecticide (parathion) absorption. The biomarker for absorption was taken to be the lower limits given by Wolfsie and Winter (1952) for RBChE depression, less than 0.53ΔpH units/h, or for PChE depression, less than 0.44ΔpH units/h. Two groups of orchard workers were actively engaged in peach picking during the same period. The reported data were used to derive the estimates of sensitivity, specificity, biomarker frequency, and event frequency shown in Table 8-3.

Because the work being done by the two picker populations was very similar, you might assume that sensitivity and specificity from one group might be applicable to the other group. However, as shown in Table 8-3, the possible ranges of sensitivity and specificity of the two groups have widely disjoint regions. The sensitivity observed in population 2 is not compatible with the allowed values in population 1. Further, the observed specificity in population 1 is not an allowable value in population 2, and that observed in population 1 is not possible in population 2. **These data emphasize clearly the true implication of understanding the meaning of sensitivity and specificity in field studies. That is, sensitivity, specificity, biomarker frequency, and event frequency must be measured on the population in which they are to be used.** Any mixture of populations in the study group on which these parameters are based will quite likely render them inapplicable to any specific population.

Once the sensitivity and specificity of a biomarker are adequately established in a population, the biomarker can be considered for use in epidemiologic studies. In certain circumstances these estimates can be used, much as the estimates of laboratory sensitivity and specificity were used earlier. Feasibility and acceptability are also necessary criteria for use, and their application will be discussed later in this chapter.

It is clear that the temporal relationship of the event and the biomarker determines the study design to be used. The underlying temporal structure is **exposure-biomarker-effect.** Clearly, the biomarker cannot precede exposure; however, the biomarker may persist indefinitely after exposure ceases, or it may be concurrent with exposure and disappear after exposure ceases. If the biomarker persists, with the event following the initial presence of the

TABLE 8-3. Illustration of the Nongeneralizability of Biomarker Sensitivity and Specificity across Populations.

	Population 1	Population 2
Sensitivity* (b)	.86	.25
Specificity* (a)	.24	.59
Biomarker frequency* (m)	.79	.36
Event frequency* (p)	.30	.36
Possible range of sensitivity (b)	$(.49,1)^\dagger$	$(0,1)^\dagger$
Possible range of specificity (a)	$(0,.30)^\dagger$	$(.56,1)^\dagger$

* Calculated from symptomatic and asymptomatic workers, Tables 1 and 2, Milby et al.
†See equations [2] and [3] in text.

biomarker by an indeterminate latent period, then a prospective or historical cohort approach can be used. For example, concurrently, two groups of people, one with and one without the biomarker, can be followed for an extended period of time until the event is measurable. This should concentrate the event in the biomarker group and dilute it in the nonbiomarker group. Or, if the historical cohort approach is used, it would be possible to use case-history data or archived samples in a well-characterized exposed population where the biomarker exists, and trace the group through time until a measurable outcome is possible. You could also use a retrospective cohort approach, using case history or archived data, where biomarker and event are well characterized, for comparison with an appropriate control group. If the biomarker is transient, with the event following at some indeterminate future point, independent samples could be used to identify prospective cohorts for follow-up. If the biomarker and event are concurrent (that is, both the biomarker and the event are present), then the cross-sectional approach may be used; however, in using this approach, we do not know whether the biomarker precedes the event or follows it. For a more complete discussion of sensitivity and specificity relating to the use of biomarkers see Griffith et al. (1989).

SUMMARY

There are many statistical and design considerations that will need to be addressed as biomarkers are put into use in epidemiologic studies. For example, misclassification error involving biomarker and event status will warrant consideration (White 1986). Epidemiologists frequently must be concerned about the need for study populations of sufficient size to permit the demonstration of health effects with a given degree of power. However, several investigators (Kraus et al. 1981; Neutra 1983) have reported the difficulties involved in getting workers to provide body tissues and fluids for sampling purposes. Thus, one might anticipate poor acceptability on the part of human subjects regarding biomarker studies. Because health professionals may view the procedures (for example, drawing blood, sample preparation and transportation, laboratory analysis) as invasive and costly, there may be additional reluctance to pursue epidemiologic studies focusing on biomarker identification and analysis.

In designing and interpreting epidemiologic studies involving the use of biomarkers, feasibility is also of some importance. Because it is unusual for an adverse biologic effect to result from a specific toxic exposure, study designs must be sufficiently broad to permit the investigation of confounding variables resulting from multichemical exposures (Goyer 1983). When a biomarker is to be used to determine individual exposure, data on variation in absorption, metabolism, pharmacokinetics, and distribution among individuals must be considered (Bloom 1981). Data are also needed on activation, detoxification, excretion, binding, replication, and repair of biomarkers (Haseltine et al. 1983).

There are also serious ethical and legal considerations in the use of biomarkers. As long as samples are from archived tissue banks (for example, blood, autopsy, fetal wastage), and individual donors are unknown to researchers, ethical and legal considerations are of less importance. However, when living donors are involved in sample collection, and it becomes necessary to apply informed consent procedures, investigators have the moral and ethical obligation to explain the purposes of the study and provide a reasonable explanation of the findings. Wolff (1985) has suggested that damage to genetic material has not yet been shown to be predictive of ill health. Consequently, researchers must be prepared to consider explanations of findings to study participants with elevated biomarkers (for example, SCEs, chromosomal aberrations, or DNA adducts), concerning the likelihood of disease development.

One option is that the investigator present the finding as biomarkers of exposure, with little or no explanation as to the possibility of disease, and that may satisfy the study participants. Perhaps as suggested by Schulte (1987), biomarkers should not be used in epidemiologic studies until validation studies, including determinations of sensitivity, specificity, and predictive value, are completed. Validation studies should emphasize the **normal population range** of the findings, the desirability of an adequate sample size, and the appropriate control of confounders. In addition, persistence of the biomarker in terms of recent or past exposure, as well as a plausible explanation for the biological mechanism involved in the suspect event, should be determined through laboratory or clinically designed experimental studies prior to undertaking epidemiologic studies. Although it is clear that there is great interest in the scientific community in applying the technology of molecular biology to epidemiologic studies, it is equally clear that the interpretation of findings has proved difficult. Consequently, guidelines should be developed by scientists, in conjunction with the medical community and other interested groups, concerning the use and interpretation of biomarkers in epidemiologic studies.

In conclusion, biomarker studies require biological samples from individuals, in contrast to the more traditional epidemiologic approach, which uses group membership to define study groups. As population distributions of biomarkers become known, then more traditional population-based studies can be developed. However, because of the potential difficulties in developing appropriate study designs and identifying willing study populations, the feasibility of biomarkers in general, and molecular biomarkers specifically, must be evaluated through considerable methodological research before they are ready for application in epidemiologic studies.

ASSIGNMENTS

1. Distinguish between biomonitoring and biological marker.
2. Discuss an appropriate setting for using each: markers of exposure, effect, and susceptibility.
3. Name some of the positive, and some of the negative factors involved in the use of biological markers in epidemiologic studies.
4. What is the relationship of biomarkers as represented in Figure 8-5, compared with levels of prevention (Figure 1-1) from Chapter 1?

REFERENCES
Albertini, R. J. 1990. "Human Environmental Monitoring Using *in vivo* Somatic Cell Gene Mutations as Indicators of Adverse Effects." Symposium lecture, 21st Annual Meeting of the Environmental Mutagen Society, Albuquerque, NM.
Allen, J. W., Sharief, Y., Langenbach, R., Waters, M. D. 1983. "Tissue-Specific Sister Chromatid Exchange Analyses in Mutagen-Carcinogen Exposed Animals." pp. 451–72, In *Organ and Species Specificity in Chemical Carcinogenesis.* eds R. Langenbach, S. Nesnow, and J. M. Rice, New York: Plenum Publishing Corporation.
Benditt, E.P. 1977. "The Origin of Atherosclerosis." *Scientific American* 236:74–85.
Bhatnagar, V. K., Saigal, S., Singh, S. P., Khemani, L. D., Malviya, A. N. 1982. "Survey Amongst Workers in Pesticide Factories." *Toxicology Letters* 10:129–32.
Bishop, J. M. 1987. "The Molecular Genetics of Cancer." *Science* 235:305–11.
Bloom, A. D. 1981. *Guidelines for Studies of Human Populations Exposed to Mutagenic and Reproductive Hazards.* White Plains, NY: March of Dimes Birth Defects Foundation, 5–163.
Boveri T. 1914. "Zur Frage der Erstehung Maligner Tumoren." *Jena, Fischer.*
Brandt-Rauf, P. W. 1991. "Oncogene Proteins as Biomarkers in the Molecular Epidemiology of Occupational Carcinogenesis." *Occupational and Environmental Health* 63:1–8.
Brilliant, L. B., Wilcox, K., Van Amburg, G., Eyster, J., Isbister, J., Bloomer, A. W., Humphrey, H., Price, H. 1978. "Breast-Milk Monitoring to Measure Michigan's Contamination With Polybrominated Biphenyls." *Lancet* ii: 643–46.
Brusick, D. 1980. *Principles of Genetic Toxicology.* New York: Plenum Press.
CDC/ATSDR. 1990. Subcommittee on Biomarkers of Organ Damage and Dysfunction. *Biomarkers of Organ Damage or Dysfunction for Renal, Hepatobiliary and Immune Systems, Summary Report.* Atlanta, GA, Centers for Disease Control/Agency for Toxic Substances and Disease Registry, 15 pp.
Clarkson, T. W., Weiss, B., and Cox, C. 1983. "Public Health Consequences of Heavy Metals in Dump Sites." *Environmental Health Perspectives* 48:113–27.
Coye, M. J., Lowe, J. A., and Maddy, K. T. 1986. "Biological Monitoring of Agricul-

tural Workers Exposed to Pesticides. I. Cholinesterase Determinations." *J. Occupational Medicine* 28:619–27.

Ehrenberg, L., Moustacchi, E., Osterman-Golkar, S., Ekman G. 1983. "Dosimetry of Genotoxic Agents and Dose/Response Relationship of Their Effects." *Mutation Research* 123:121–82.

Everson, R. B., Randerath, E., Santella, R. M., Cefalo, R. C., Avitts, T. A., Randerath, K. 1986. "Detection of Smoking-Related Covalent DNA Adducts in Human Placenta." *Science* 231:54–57.

Garry, V. F., Griffith, J., Danzl, T. J., Nelson, R. L., Whorton, E. B., Krueger, L. A., Cervenka, J. 1989. "Human Genotoxicity: Pesticide Applicators and Phosphine." *Science* 246; 251–55.

Goyer, R. A. 1983. "Introduction and Overview." *Environmental Health Perspectives* 48:1–2.

Griffith, J., and Duncan, R. C. 1985. "Alkyl Phosphate Residue Values in the Urine of Florida Citrus Fieldworkers Compared to the National Health and Nutrition Study (HANES) Sample." *Bulletin of Environmental Contamination and Toxicology* 34:210–15.

Griffith, J., Duncan, R. C., and Hulka, B. S. 1989. "Biochemical and Biological Markers: Implications for Epidemiologic Studies." *Archives of Environmental Health* 44:375–81.

Grobler, S. R., Maresky, L. S., Rossouw, R. J. 1986. "Blood Lead Levels of South African Long Distance Runners." *Archives of Environmental Health* 41:155–58.

Harlan, W. R., Landis, J. R., Schmouder, R. L., Goldstein, N. G., Harlan, L. C. 1985. "Blood Lead and Blood Pressure Relationship in the Adolescent and Adult U.S. Population." *JAMA* 253:530–34.

Hartman, P. E. 1983. "Mutagens: Some Possible Health Impacts Beyond Carcinogenesis." *Environmental Mutagen* 5:139–52.

Harris, C. C., Autrup, H., Stoner, G. 1978. "Metabolism of Benzo(a) Pyrene in Cultured Human Tissues and Cells." pp. 331-342, In: *Polycyclic Hydrocarbons and Cancer, Vol 2,* eds G. Stoner, H.V. Gelboin, and P.O. T'so. New York: Academic Press.

Haseltine, W. A., Franklin, W., Lippke, J. A. 1983. "New Methods for Detection of Low Levels of DNA Damage in Human Populations." *Environmental Health Perspectives* 48:29–41.

Heath, C. W., jr. 1983. "Field Epidemiologic Studies of Populations Exposed to Waste Dumps." *Environmental Health Perspectives* 48:3–7.

Heddle, J. A., Hite, M., Kirkhart, B., Mavournin, K., MacGregor, J. T., Newell, G. W., Salamone, M. F. 1983. "The Induction of Micronuclei as a Measure of Genotoxicity." *Mutation Research* 123:61–118.

Henderson, R. F., Bechtold, W. E., Bond, J. A., and Sun. J. D. 1989. "The Use of Biological Markers in Toxicology." *Critical Reviews of Toxicology* 20:65–82.

Hulka, B. S. 1990. "Overview of Biological Markers." pp. 3–15, In: B.S Hulka, T.C. Wilcosky, and J.D. Griffith eds In *Biological Markers in Epidemiology*. New York: Oxford University Press.

Jauhar, P. P., Henika, P. R., MacGregor, J. T., Wehr, C. M., Shelby M. D., Murphy, S. A., Margolin, B. H. 1988. "1,3-butadiene: Induction of Micronucleated Eryth-

rocytes in the Peripheral Blood of B6C3F1 Mice Exposed by Inhalation for 13 Weeks." *Mutation Research* 209:171–76.
Khoury, M. J., Newill, C. A., and Chase, G. A. 1985. "Epidemiological Evaluation of Screening for Risk Factors: Application to Genetic Screening." *American Journal of Public Health* 75:1204–08.
Kraus, J. F., Mull, R., Kurtz, P., Winterlin, W., Franti, C. E., Borhani, N., Kilgore, W. 1981. "Epidemiologic Study of Physiological Effects in Usual and Volunteer Citrus Workers From Organophosphate Pesticide Residues at Reentry." *Journal of Toxicology and Environmental Health* 8:169–84.
Leavell, B. S., Thorup, O. A. 1976. *Fundamentals of Clinical Hematology*, 4th ed. Philadelphia: W. B. Saunders Co.
Lower, G. M. 1982. "Concepts in Causality: Chemically Induced Human Urinary Bladder Cancer." *Cancer* 49:1056–66.
Loveless, A. 1969. "Possible Relevance of 0-6 Alkylation of Deoxyguanosine to the Mutagenicity and Carcinogenicity of Nitrosamines and Nitrosamides." *Nature* 223:206–7.
McLaughlin, M., Linch, A. L., Snee, R. D. 1983. "Longitudinal Studies of Lead Levels in a US Population." *Archives of Environmental Health* 27:305–11.
Milby, T., Ottoboni, H., and Mitchell, H. W. 1964. "Parathion Residue Poisoning Among Orchard Workers." *JAMA* 189:351–56.
Miller, S. 1984. "Biological Monitoring." *Environmental Science and Technology* 18:188A–190A.
Neutra, R. 1983. "Roles for Epidemiology: The Impact of Environmental Chemicals." *Environmental Health Perspectives* 48:99–104.
NRC. 1983. *Identifying and Estimating the Genetic Impact of Chemical Mutagens.* Washington, D.C., National Academy Press, 295 pp.
NRC. 1987. "Biological Markers in Environmental Health Research." *Environmental Health Perspectives* 74:3–9.
NRC. 1989a. *Biological Markers in Pulmonary Toxicology.* Washington, D.C., National Academy Press, 179 pp.
NRC. 1989b. *Biological Markers in Reproductive Toxicology.* Washington, D.C., National Academy Press 395 pp.
Omenn, G. S. 1982. "Predictive Identification of Hypersusceptible Individuals." *Journal of Occupational Medicine* 24:369–74.
Perry, P., and Evans, H. J. 1975. "Cytological Detection of Mutagen-Carcinogen Exposure by Sister Chromatid Exchange." *Nature* 258:121–25.
Perera, F., Mayer, J., Santella, R. M., Brenner, D., Jeffrey, A., Latriano, L., Smith, S., Warburton, D., Young, T. L., Tsai, W. Y., Hemminki, K., Brandt-Rauf, P. 1991. "Biological Markers in Risk Assessment for Environmental Carcinogens." *Environmental Health Perspectives* 90:247–54.
Perera, F. P., Santella, R. M., Brenner, D., Poirier, M. C., Munshi, A. A., Fischman, H. K., Van Ryzin, J. 1987. "DNA Adducts, Protein Adducts and Sister Chromatid Exchange in Cigarette Smokers and Non-smokers." *JNCI* 79:449–56.
Perera, F. P., and Weinstein, I. B. 1982. "Molecular Epidemiology and Carcinogen-DNA Adduct Detection: New Approaches to Studies of Human Cancer Causation." *Journal of Chronic Disease* 35:581–600.

Pirkle, J. L., Schwartz, J., Landis, J. R., Harlan, W. R. 1985. "The Relationship Between Blood Lead Levels and Blood Pressure and Its Cardiovascular Risk Implications." *American Journal of Epidemiology* 121:246–58.

Rothwell, C. J., Hamilton, C. B., Leaverton, P. E. 1991. "Identification of Sentinel Health Events as Indicators of Environmental Contamination." *Environmental Health Perspectives* 94:261–63.

Schnell, F. C., and Chiang, T. C. 1990. *Protein Adduct Forming Chemicals for Exposure Monitoring: Literature Summary and Recommendations.* U.S. Environmental Protection Agency, EPA/600/4-90/007, 134 pp.

Schulte, P. A. 1987. "Methodologic Issues in the Use of Biologic Markers in Epidemiologic Research." *American Journal of Epidemiology* 126:1006–16.

Schwartz, G. G. 1990. "Chromosome Aberrations." pp. 147–72, In: *Biological Markers in Epidemiology*, eds. B. S. Hulka, T. C. Wilcosky, and J. D. Griffith. New York: Oxford University Press.

Seeger, R. C., Brodeuk, G. M., Sather, H., Dalton, A., Siegal, S., Wong, K. Y., Hammond, D. 1985. "Association of Multiple Copies of the N-myc Oncogene With Rapid Progression of Neuroblastomas." *New England Journal of Medicine* 313:1111–16.

Seidegard, J., Pero, R. W., Miller, D. G., Beattie, E. J. 1986. "A Glutathione Transferase in Human Leukocytes as a Marker for the Susceptibility to Lung Cancer." *Carcinogenesis* 7:751–53.

Singh, N. P., Danner, D. B., Tice, R. R., Brant, L., Schneider, E. L. 1990. "DNA Damage and Repair With Age in Individual Human Lymphocytes." *Mutation Research* 237:123–30.

Singh, N. P., McCoy, M. T., Tice, R. R., Schneider, E. L. 1988. "A Simple Technique for Quantitation of Low Levels of DNA Damage in Individual Cells." *Exp Cell Res* 175:184–91.

Stock, L. M., Brosman, S. A., Fehey, J. L., Liu, B. 1987. "C-S Ras Related Oncogenic Protein as a Tumor Marker in Transitional Cell Carcinoma of the Bladder." *Journal of Urology* 137:7889–92.

Stolley, P. D., Soper, K. A., Galloway, S. M., Nichols, Norman, S. A., Wolman, S. R. 1984. "Sister Chromatid Exchanges in Association With Occupational Exposure to Ethylene Oxide." *Mutation Research* 129:89–102.

Taylor, H. J. 1958. "Sister Chromatid Exchanges in Tritium-Labeled Chromosomes." *Genetics* 43:515–29.

Tice, R. R. 1990. *Evaluation of Exposure Markers.* EPA/600/4-90/034. 95 pp.

U.S. EPA. 1990a. *Chromosomal Aberration Data Analysis and Interpretation System: User's Guide and Software.* EPA/600/8-90/086. 94 pp.

U.S. EPA. 1990b. *Micronucleus Assay Data Management and Analysis System: User's Guide and Software.* EPA/600/4-90/011. 35 pp.

Vine, M. F., and McFarland, L. T. 1990. "Markers of Susceptibility." pp. 196-213, In: *Biological Markers in Epidemiology*, eds. B. S. Hulka, T. C. Wilcosky, and J. D. Griffith. New York: Oxford University Press.

White, E. 1986. "The Effect of Misclassification of Disease Status in Follow-up Studies: Implications for Selecting Disease Classification Criteria." *American Journal of Epidemiology* 124:816–25.

Wicker, G. W., Williams, W. A., Gutherie, F. E. 1979. "Exposure of Fieldworkers to Organophosphorus Insecticides: Sweet Corn and Peaches." *Archives of Environmental Contamination and Toxicology* 8:175–82.

Wickizer, T. M., Brilliant, L. B., Copeland, R., Tilden, R. 1981. "Polychlorinated Biphenyl Contamination of Nursing Mothers' Milk in Michigan." *American Journal of Public Health* 71:132–37.

Wiencke, J. K., Kelsey, K. T., Lamela, R. A., Toscano, W. A. Jr. 1990. "Human Glutathione S-Transferase Deficiency as a Marker of Susceptibility to Epoxide-Induced Cytogenetic Damage." *Cancer Research* 50:1585–90.

Wilcosky, T. C. and Griffith, J. D. 1990. "Applications of Biological Markers." pp. 16–27, In: *Biological Markers in Epidemiology*, eds. B. S. Hulka, T. C. Wilcosky, and J. D. Griffith. New York: Oxford University Press.

Wilcosky, T. C., and Rynard, S. M. 1990. "Sister Chromatid Exchanges." pp. 105–124, In: *Biological Markers in Epidemiology*, eds. B. S. Hulka, T. C. Wilcosky, and J. D. Griffith. New York: Oxford University Press.

Wolff, S. 1985. "Problems and Prospects in the Utilization of Cytogenetics to Estimate Exposure at Toxic Chemical Waste Dumps." *Environmental Health Perspectives* 48:25–27.

Wolfsie, J. H., and Winter, G. D. 1952. "Statistical Analysis of Normal Human Red Blood Cell and Plasma Cholinesterase Activity Values." *Archives of Industrial Hygiene and Occupational Medicine* 6:43–49.

Zarbl, H., Sukumar, S., Arthur, Av., Martin-Zanca, D., Barbaci, M. 1985. "Direct Mutagenesis of Ha-ras-1 Oncogenes by N-nitroso-N-methylurea During Initiation of Mammary Carcinogenesis in Rats." *Nature* 315:382–85.

9

Disease and the Environment

Tim Aldrich, Jack Griffith, and Robert Meyer

OBJECTIVES

This chapter will:

1. Discuss the public's perception of disease occurrence.
2. Review the role of the media in developing public opinion.
3. Introduce the three classes of disease endpoints that are commonly related to environmental exposures to toxic chemicals.

PUBLIC PERCEPTION

Cardiovascular disease (diseases of the heart and blood vessels) is the leading cause of death in both men and women. Interestingly, since 1965, there has been more than a 40 percent reduction in mortality due to ischemic heart disease. Much of this reduction in mortality has occurred since 1982, and is taking place in many parts of the world, including the United States. For example, age-standardized rates for ischemic heart disease among men and women of the Hunter Region of Australia are shown in Figure 9-1. Men experienced a decline in disease of about 3.7 percent (S.E. ± 1.6) per year, while the women experienced a reduction of almost 5 percent (4.9 ± 2). This decline in ischemic heart disease mortality has also been recognized in the United States since the late 1960s, and Kaplan et al. (1988) reported on a 45 percent reduction of the disease in Alameda County, California. The data in Figure 9-2 reflect both a statistically significant ($p < 0.03$) decline in ischemic

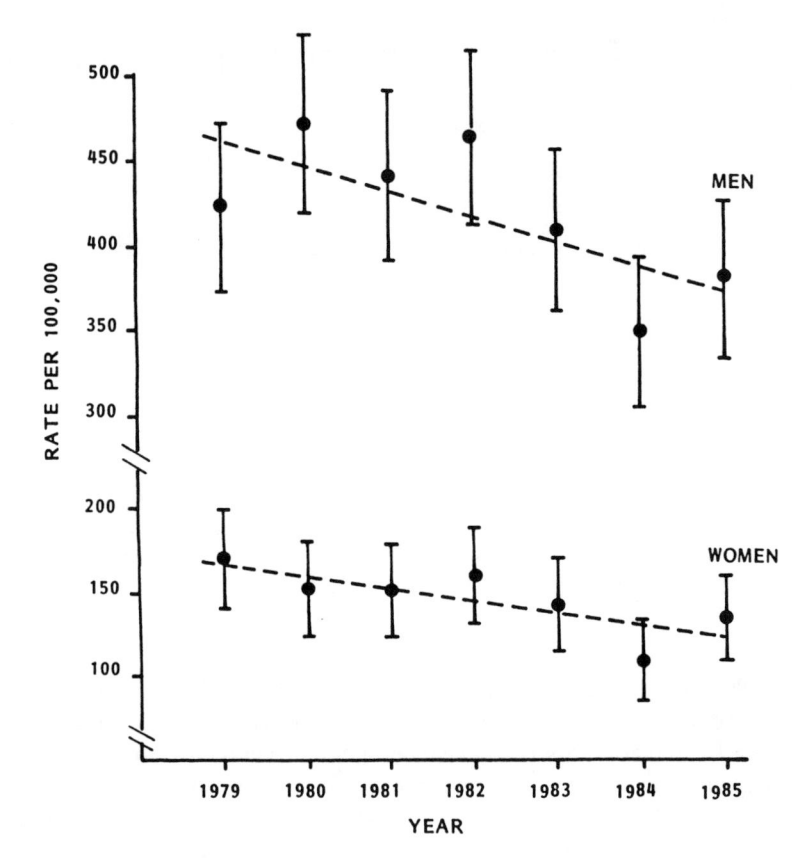

FIGURE 9-1. Mortality Rates per 100,000 From Ischemic Heart Disease, for Australian Men and Women of the Hunter Region of New South Wales, 1979–1985 (Age-Standardized Rates ± Two Standard Deviations). Source: Adapted from Dobson et al. 1988.

heart disease and the use of preventive measures **(for example, weight loss, exercise, and cessation of smoking)** among the cohorts.

When injury, and other noncancer causes of death are counted with cardiovascular mortality, these diseases remain responsible for a majority of all deaths in the United States. For example, the graph in Figure 9-3, suggests that the risk of death is much higher for males and females in noncancer diseases compared with deaths in both sexes from respiratory and non-respiratory cancers.

Although it is well understood in the public health community that cardiovascular disease is the major cause of death in the United States and Europe, and that atherosclerosis—the primary cause of myocardial and cerebral infarction—is responsible for a majority of these deaths (Williams

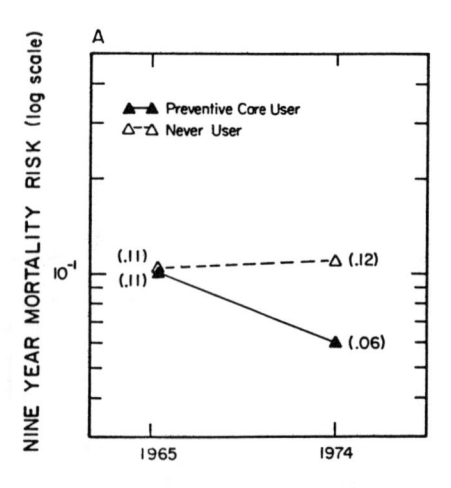

FIGURE 9-2. Nine-Year Ischemic Heart Disease Mortality Risk in the 1965 and 1974 Cohorts of the Alameda County Study by Preventive Care Use Estimated by a Logistic Model Containing Terms for Age, Sex, Race, Cohort, Preventive Care Use, and Cohort × Preventive Care Use Evaluated for White Males Age 60. Source: Cohn et al. 1988.

et al. 1989), the public perception is that cancer (arguably the most media-promoted disease in the United States, Allman 1985) is the greater risk. While it appears from media reports that cancer incidence and mortality are on the upswing, with few exceptions (for example, female breast cancer) this may not be the case. In fact, the appearance of relatively more cancer may reflect the success that the medical and health professionals are having in preventing and curing other diseases (Doll and Peto 1981). However, the attention of the public is decidedly focused on cancer, a condition that is presented to the public as a long-term, highly debilitating disease, ending almost inexorably in death.

As a result of media exposure, cancerphobia and carcinogen of the week have become popular catch phrases. Both terms, in part reflect the public's confusion over what they are told in the media, by their physicians, and even by friends. It is important, however, that health professionals understand that media interest has its positive aspects. Media promotions for cancer screening aid in case finding through early detection and treatment and are important factors involved in attempting to find a cure for this deadly disease. Finally, we should recognize that the public's perception of the health risks associated with certain exposures does not necessarily lead to preventive action (for example, the risk of smoking and lung cancer is well known, yet millions of Americans continue to smoke). With AIDS, the public is certainly confused as to whether the perceived risk is **voluntary** (transmitted by sex with con-

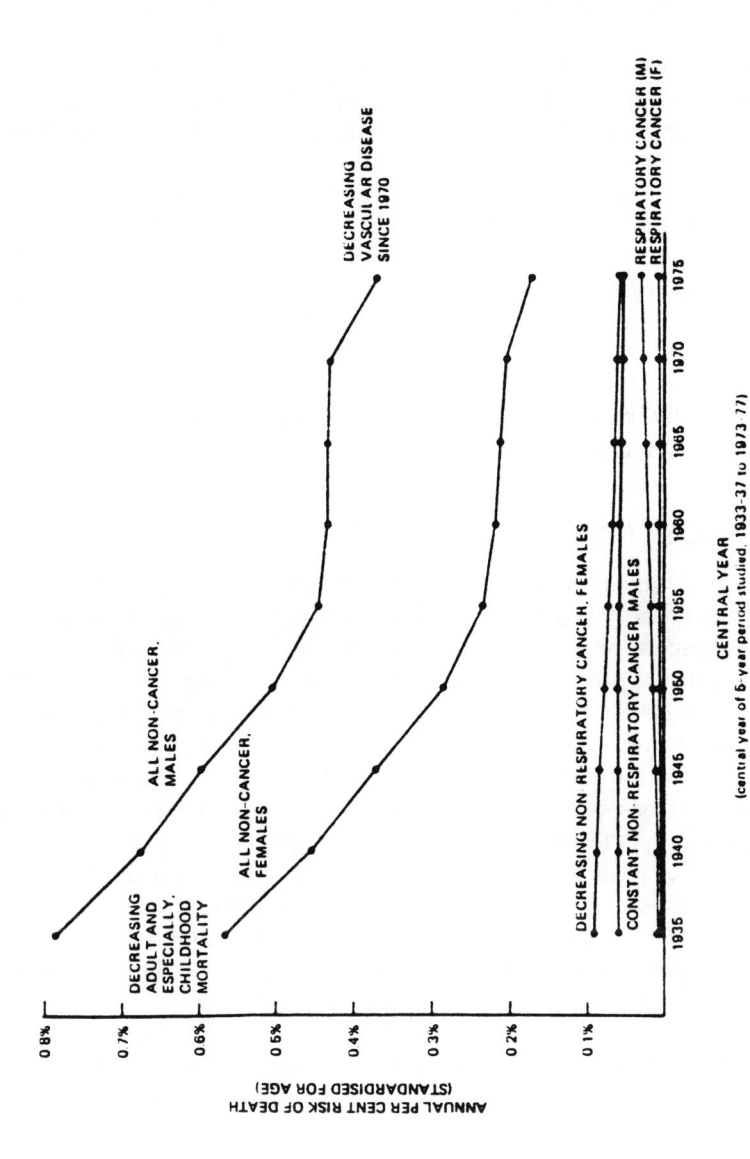

FIGURE 9-3. Annual Age-Standardized Death Rates, 1933–1977, Among Americans Under 65 Years of Age. Source: Doll and Peto 1981.

senting adults), or involuntary (for example, blood transfusions, poorly sterilized medical equipment, or casual intimacy, such as kissing, holding hands, drinking after someone with AIDs) (Freudenberg 1988; Schechter et al. 1989; Howe 1988).

Clearly, when a person is able to assume responsibility for his or her actions (for example, wearing a seatbelt, wearing a condom), there will be less fear and less anxiety. In the following text, we are going to briefly discuss carcinogenesis, reproductive effects, and neurotoxic effects, as these health endpoints relate to the application of environmental epidemiologic methods. Since much of the ongoing interest in environmental epidemiology focuses on cancer, that will also be the primary focus of our discussion. But, the reader should not construe from this that we place less importance on the remaining endpoints. Reproductive and neurologic effects play a significant role in understanding environmentally related toxic exposures. Unfortunately, less emphasis is placed on these outcomes in terms of the public's perception of environmentally related health risks, resulting in a corresponding reduction in available dollars for research.

CARCINOGENESIS

Carcinogenesis involves the production of a carcinoma: A malignant new growth from the epithelial tissue that invades surrounding tissue and spreads from one organ site or tissue to another, a process known as metastases (Marx 1989). The more appropriate term for the class of disease we collectively refer to as cancer is carcinoma (from the greek *karkinos*, meaning crab). Cancer is an ancient term arising from the visual impression of a surgeon that a tumor appeared like a crab with legs extending into the surrounding tissue.

Although it has been understood for some time that the process of carcinogenesis is molecular (occurring at the cellular level), only through recent advances in molecular biology are we beginning to understand the process. Mollgavkar (1983) first proposed a two-step process for carcinogenesis, and coined the terms **initiation** and **promotion**. Initiation is the transformation of a benign or harmless cell to one with the potential for malignant growth. Initiation is, however, distinct from the event that causes a cell to grow rapidly and to acquire the capacity to metastasize. Unrestrained growth is believed to follow another distinct event: promotion (Shubik 1984).

The concepts of initiation and promotion are far too simplistic for understanding actual carcinogenesis. Recently, discovery of oncogenes (that is, genetic material passed from parent to offspring with the potential for causing cancer) and the genetic mapping of fragile sites on the **genome** (the complete set of hereditary factors), have enhanced the general understanding of carcinogenesis. It is hypothesized that carcinogenesis requires a series of cellular events to occur, and that some of these events may require a specific sequence to be effective.

For example, with familial breast cancer (Hall et al. 1990), and possibly with colon cancer (Kinzler et al. 1991), specific genes have been located on the genome that must be activated in addition to a subsequent loss of regulation for those genes. In both cases genetic changes favoring cell survival and growth are necessary adjuncts to the disease process, although the exact sequence of these sustaining events may not be as critical as with those involving the key genetic locus. Often, nutritional factors interact with these genetic events to provide a favorable potential for growth.

For the purposes of environmental epidemiology it is necessary to recognize the genetic–cellular nature of carcinogenesis and the interplay of multiple forces in order to achieve carcinogenesis. Carcinogenesis may occur at the genetic–cellular level through many means. For example, although the role viruses play in carcinogenesis is not well understood, with several cancers (for example, cervical cancer, liver cancer, nasopharyngeal cancer, and Hodgkin's disease) a role for a viral agent is well accepted.

Clastogenesis literally means chromosome damage. Chromosome damage takes the shape of deletions, breaks, gaps, rings, dicentrics, quadriradial figures, and acentric fragments. Gaps are defined as achromatic regions within a chromatid less than the width of the chromatid. Deletions in the nonbanded preparation are nonstaining regions in a chromatid greater than the width of the chromatid. Breaks are a discontinuity in a chromatid or chromosome that is misaligned. Rings, dicentrics, quadriradial figures, and acentric fragments result from exchanges within or between chromatids. Similarly, mutational events, whether additions or deletions, are also a well-accepted means for genetic transformation potentially relevant to cancer risk (initiation).

Environmental agents may now be designated as mutagens, clastogens, and mitogens (for example, stimulating cell division). To these cell-damaging ideas must be added the topics of cellular repair and regulation (Weinstein 1991). Although cells are quite proficient at genetic repair, repair capabilities vary in human populations. Also varying between people is the ability to detoxify chemicals that enter the body. The liver acts as a first-line defense for chemical hazards. The body prefers to remove or detoxify a hazardous chemical by combining it with glutathione (a chemical that is easily recognized and excreted) in a process known as conjugation. However, there are limits to the body's store of glutathione and so alternative pathways for detoxification are required. Some of these are less efficient than others, so that errors may be made. Such errors may result in a chemical becoming more (not less) toxic as it metabolizes through the system.

A good example of such an error is in the activation of vinyl chloride as it metabolizes. When an alternative detoxification process to glutathione conjugation is involved with a vinyl chloride molecule, enzymes begin the process of breaking down the vinyl chloride molecule. However, because this process is inefficient, the breaking down process may be incomplete and the partially metabolized vinyl chloride molecule remains as a hyperreactive epoxide (a much more dangerous agent). The epoxide seeks a chemical to stabilize its reactivity and will bind readily with DNA. This DNA bonding is a genetic level event that can produce a mutation (that is, the first step toward carcinogenesis). Thus there are new terms for these "pieces of the puzzle." Vinyl chloride is considered a procarcinogen since it requires an action by the error-prone detoxification process to become a proximal carcinogen.

There are similar terms for the extent to which an agent can both transform cells and promote cell growth. Agents may be termed initiators or promoters depending on their mode of action at the cellular level. Some agents have the capacity for both activities (for example, cigarette smoke) so these may be termed complete carcinogens.

Some agents are considered sufficient causes of cancer, meaning they can produce carcinogenesis without interaction with other agents (for example ionizing radiation). Some agents are necessary carcinogens, meaning that a particular cancer cannot occur without this agent being present (for example, Epstein-Barr virus in Hodgkin's disease). When you consider cancer as an outgrowth of an exposure to an environmental agent, you should also understand:

1. The pathway of exposure (see Chapter 6, "Exposure Assessment") because of the implications for activation of a potential procarcinogen.
2. Individual susceptibility since it may impact on the detoxification method that will be evoked with a particular exposure.
3. Toxicological concepts (that is, interaction of chemicals and competition for cellular receptors [Murphy 1980]).

An exemplary reference for these myriad concepts is shown in Figure 9-4. It contains not only sources but the many biologic forces that are part of a person's response to exposure to a chemical hazard (Ballantyne 1985). These issues are not made simpler by the proposition that we are never exposed to a single agent from the environment. Exposures are always complex mixtures, whether due to multiple agents in the air, soil, or water or from a more direct source (that is, diet). Thus we see that understanding carcinogenesis is incredibly complex.

If we accept cancer as a given, there is still much to learn about the diseases (cancer is not simply one disease). A common adage is that cancer is over 100 diseases (Cancer Facts and Figures 1991). This perspective arises

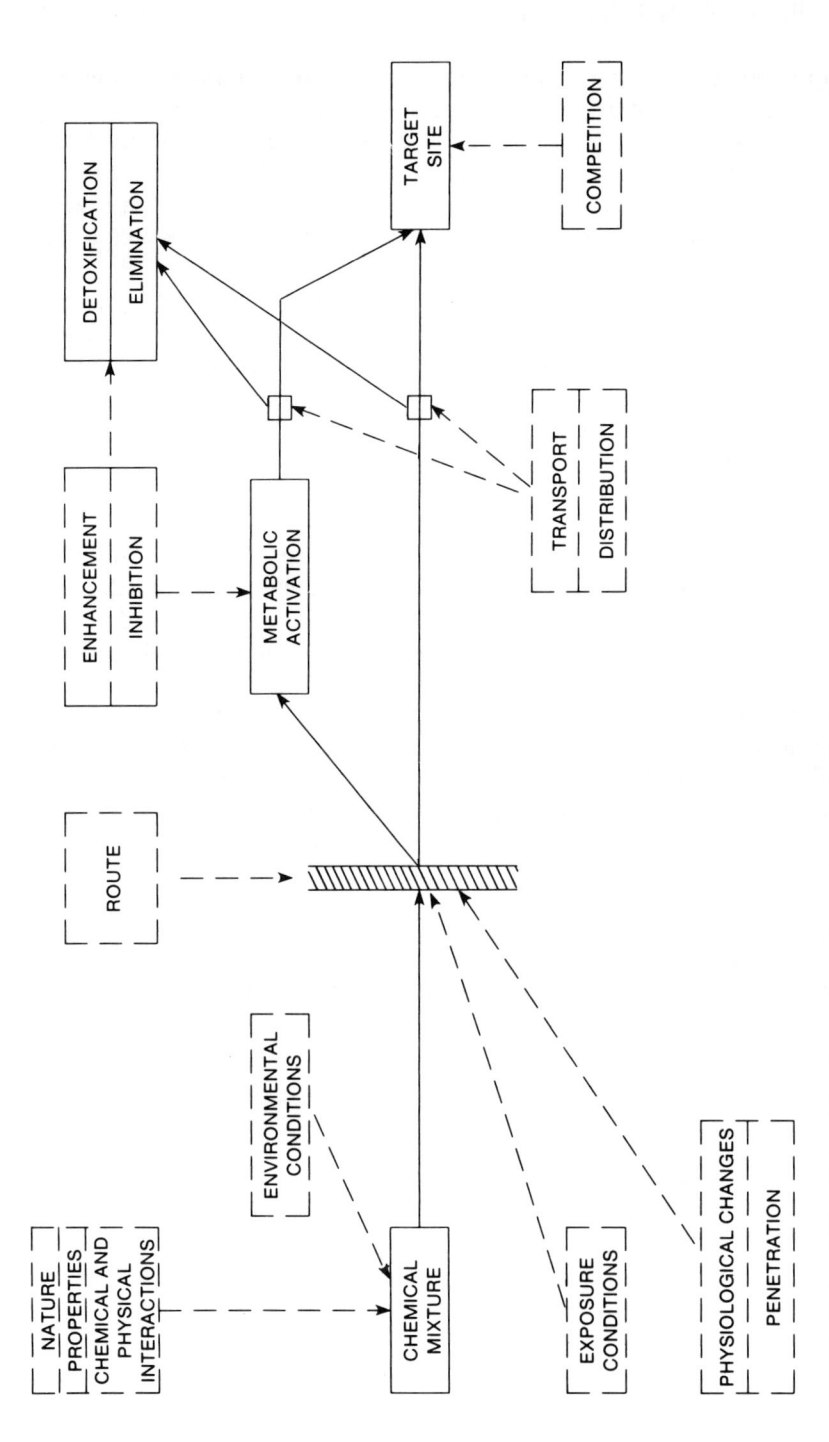

FIGURE 9-4. Major Factors That Act as Determinants for Nature, Likelihood, and Severity of Adverse Health Effects Occurring as a Result of Exposure to Mixtures of Chemicals. Continuous Lines Represent Biological Handling of Materials; and Interrupted Lines, Determinants for Expression of Toxicity. Source: Ballentine 1985.

from the 40 or so anatomic sites where cancer strikes and the various cell types that are present at many of these sites. Some cancers are more fully studied than others; these are generally the more common cancers. Some cancers are considered to have relatively well-established etiologies (etiology is the study of causation). In the next section, we will discuss several cancers in the perspective of well-documented or strongly suspected risk factors.

Cancer is an historic disease, having been observed in ancient Egypt and Greece (Shimkin 1977). Even dinosaur skeletons have been discovered with cancers. However, cancer emerged as a major public health concern in the middle of this century. The emergence of cancer as a leading cause of death is closely linked to the eradication of infectious diseases as major causes of death and the extension of life expectancy (see Table 9-1). Heart disease, cancer, and stroke are all diseases of aging, and as life expectancy rose, these causes of death rose as well (Bailar and Smith 1986).

Cancer rates have varied over this century (Figure 9-5). The most telling time trend is the precipitous rise in lung cancer death rates and the declines of uterine, liver, and stomach cancer mortality. A discussion of cancer epidemiology must necessarily include the "big" cancers and the "bad" cancers. When epidemiologists speak of big cancers, they refer to the most commonly occurring cancers (for example, lung cancer is the biggest cancer in terms of lives claimed). From Figure 9-5, we observe less than 10 of the 40 cancer sites mentioned previously. However, if all sites were included, the mortality lines would be near the bottom of the graph. Among the cancers shown in Figure 9-5, after 1960 there are only four major cancers: lung, breast, colo-rectum, and prostate.

TABLE 9-1. Changing Order of Leading Causes of Death During the Twentieth Century.

1910	1950	1990
Flu	Heart Disease	Heart Disease
T.B.	Cancer	Cancer
Diarrhea/Enteritis	Stroke	Stroke
Heart Disease	Accidents	Accidents
Stroke	Diseases of Infancy	Flu/Pneumonia
Nephritis	Flu/Pneumonia	Diabetes
Accidents	General Arteriosclerosis	Cirrhosis
Cancer	Diabetes	General Arteriosclerosis
Diseases of Infancy	Birth Defects	Suicide
Diphtheria	Cirrhosis	Diseases of Infancy
Meningitis	Other Circulatory Diseases	Homicides
Typhoid	Suicide	Emphysema

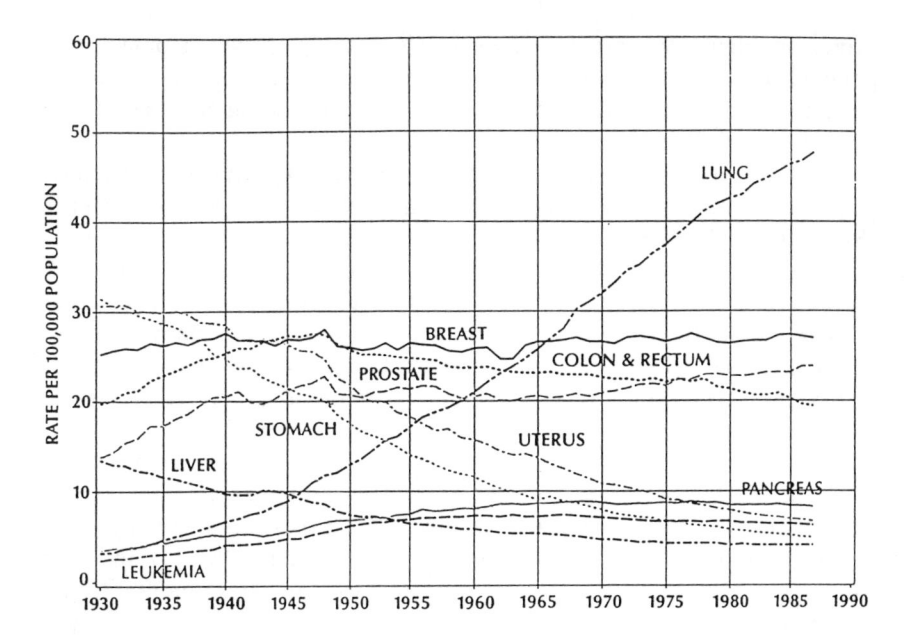

FIGURE 9-5. Rate for the Population Standardized for Age on the 1970 U.S. Population. Sources: National Center for Health Statistics and Bureau of the Census, United States. Note: Rates are for both sexes combined except breast and uterus (female population only) and prostate (male population only).

It is instructive to note that Figure 9-5 also offers another, but more subtle, perspective on cancer. The pattern of cancer for 1930–35 is much like that of the Third World today (that is, liver, stomach, and uterine cancer predominate. As countries become more urbanized/industrialized/westernized, cancer patterns shift to reflect the latter years of Figure 9-5). These changes, as well as studies of cancer patterns among migrants, are some of the strongest evidence for the assertion that most cancers arise from environmental sources (Ballantyne 1985). However, environment involves lifestyle and diet as well as the ambient surroundings.

Epidemiologists must also consider incidence as well as mortality; cancer of the prostate is the most common cancer among men, but lung cancer takes many more lives; breast cancer is more common in women, but lung cancer claims many more lives. Because of therapeutic advances, earlier detection and greater access to state-of-the art care, survivorship is increasing (that is, today, more cancer patients survive their cancer than do not). This raises the issue of the bad cancers. Bad cancers are those for which the five-year survival rates are less than 50 percent. Among the big cancers (defined from Figure 9-5 after 1960), only lung cancer meets this definition. Some other cancers shown in Figure 9-5 are bad cancers (for example, liver, stomach, and pancreas cancers, leukemia, yet these are mercifully rare). While rare and declining in the western world, however, they are leading causes of cancer death in Third World countries.

Lung cancer has come to have such an overwhelming impact that scientific publications have begun to report patterns separate from all other cancers (SEER 1989; Davis et al. 1990). The shifting pattern of cancer rates over the last three decades, in the United States and around the world, has led to considerable debate as to the reality of an overall increase in cancer rates (Bailar and Smith 1986). With the rising industrialization of many countries, the potential contribution of environmental factors to cancer etiology cannot be ignored. To better follow a discussion of environmental carcinogenesis, let us now review the general epidemiology of several of the big cancers.

Lung Cancer. Lung cancer is the leading cause of cancer death in men and women. Increased risk is strongly associated with cigarette smoking—either direct or passive exposure. Occupational exposures include arsenic, organic chemicals, asbestos, and ionizing radiation exposures (for example, radon gas). Occupational risks are increased for smokers. Vitamin A deficiency is also a suspected risk factor.

Prostate Cancer. The highest occurrence of prostate cancer is among blacks, and is primarily a disease of older men (> 65 years). Familial association and risk from dietary fat are suspected. Cadmium is a potential occupational risk.

Breast Cancer. Breast cancer is the most common and feared cancer in women. Although it is found in younger women, breast cancer is primarily a disease of post-menopausal women (> age 50). Increased risk is associated with family history and with not bearing children prior to age 30. A role for dietary fat is suspected.

Colo-Rectal Cancer. Colo-rectal cancer is the third leading cause of cancer death in both men and women. It is associated with low fiber and/or high animal fat diet, and with a history of polyps and inflammatory bowel disease.

Uterine Cancer. Cervical cancer is a disease of young women, associated with early intercourse, multiple partners, and cigarette smoking. Endometrial cancer is a disease of older women associated with infertility, estrogen therapy, and obesity.

Oral Cancer. Oral cancer is more common in men, and is associated with cigarette, cigar, and pipe smoking, smokeless tobacco use, and alcohol consumption.

Bladder Cancer. Bladder cancer is more common in whites and in men. Smoking is a recognized risk factor, and workers in dye, leather, and rubber occupations are at higher risk.

Pancreatic Cancer. Pancreatic cancer is more common in blacks, with higher rates for males and persons over age 65. Smoking is a recognized risk factor; dietary fat, chronic infections, diabetes, and cirrhosis are suspected.

Skin Cancer. Common skin cancers occur in half of all people; 98 percent survival however has led to these cancers being excluded from most statistical reports. Melanoma is the most common lethal form. Fair complexion is a strong risk factor, as is excessive exposure to the sun. Coal tar, pitch or creosote, arsenic, and radium are occupational risks.

Leukemia. Leukemia is a disease of children and older adults, both sexes, and all races. Certain genetic risks are known (for example, Down syndrome), as are viral agents (HTLV-1). Occupational risks are ionizing radiation and benzene.

Ovarian Cancer. Ovarian cancer is a disease of older women (> 60 years), and ones who have never borne children. Risk is increased by a history of breast, colo-rectal, and endometrial cancer.

Brain Cancer. The increasing occurrence in recent years may be associated with job-related aromatic hydrocarbon exposures and nonionizing radiation.

Lymphoma. Lymphoma has been increasing in recent years. This is essentially a mixed group of cancers with many suspected risk factors including agricultural chemicals, viruses, and childhood exposures.

Stomach Cancer. A leading cancer in developing countries, it is associated with nitrates in food.

Liver Cancer. A leading cancer in developing countries, associated with hepatitis B infection, cirrhosis, and occupational exposure to aromatic hydrocarbons.

In studying environmental agents, there are many considerations, but among the most important is understanding the recognized risk factors. Much of the emphasis on environmental cancer epidemiology is directed to ecologic studies. In this instance, the usefulness of standardized rates (age, race, and sex adjustment of disease rates) and expected numbers is well established. With politically visible cancer, however, the emotional impact to a community may be accentuated even when expected disease patterns are observed. At any rate, it is important to know the **expected** risk in a community study, since this knowledge will be a valuable asset in setting study priorities and in educating the public. Similarly, knowing which cancers have established risk factors, and being able to reason over the distribution of the risk factors in a community, is also useful. Epidemiologists should always review the cancer literature before undertaking a new study. Because the literature is so plentiful, with new hypotheses arising so quickly, it remains difficult to stay abreast of the most recent events.

Investigating Environmentally Related Cancers

Cancer investigations are difficult (that is, numbers of cases are usually small, and exposures confounded). From Doll and Peto's work (Table 9–2), one sees that lifestyle (for example, diet, tobacco use, alcohol consumption, and personal sexual history) is the most common risk factor for cancer (Doll and Peto, 1981). The following terms will have profound influence on any environmentally related cancer investigation:

1. **Induction.** The interval between an exposure to a carcinogen and the induction of a transformed cell line is termed **induction.** Induction is believed to require many years to occur—primarily due to the effectiveness of molecular repair mechanisms. With most occupational studies, it is not unusual to assume an induction period of five to ten years.
2. **Latency.** After induction or initiation, there is an interval that permits promotion to occur and then for overt disease to be detected. This is believed to be a much slower process than for initiation. The interval from

TABLE 9-2. Percentage of Cancer Deaths Attributable to Different Factors.

Factor	Range of Best Estimates	Estimates
Lifestyle:		
Tobacco	25–40	30
Diet	10–70	35
Infection	1	10
Reproduction and sexual behavior	1–13	7
Alcohol	2–4	3
TOTAL	39–100	85
Societal:		
Occupation	2–8	4
Industrial products	< 1–2	< 1
Food additives	2	< 1
Medicines and medicinal procedures	0.5–3	1
TOTAL	4–15	6
Environmental:		
Pollution	< 1–5	2
Geophysical factors	2–4	3
TOTAL	3–9	5
Cumulative attributable risks	46–100	96

Source: Doll and Peto, 1981.

the induction of disease until it is clinically detected is termed **latency.** For most cancers, latency is believed to be 10 to 20 years. Again, with occupational epidemiologic studies, time intervals are frequently included with case eligibility considerations to accommodate a latency interval (that is, in considering an exposure one must look back in time for as much as 30 years).

Although cancer seems to be prominently mentioned when confronting environmentally related controversies, environmental exposures may account for less than 5 percent of all cancers. For this reason, we recommend that environmental epidemiologists focus on relatively rare cancers—to avoid false leads from patterns identified by the public ("everyone in my town has cancer") and to maximize the opportunity to identify a single agent. Another advantage of studying rare cancers is that many of them are bad cancers (for example, liver and brain cancer, some of the leukemias and lymphomas), thus presenting a significant threat to the public health. Higher case-fatality may make for shorter induction and latency intervals, and with shorter disease intervals may come the advantage of better recall of exposure histories and the loss of fewer case-families to relocation.

With sparse populations and rare events, statistical artifacts become more of a **real** as opposed to **potential** problem. Thus, careful biologic reasoning is needed, complemented by subtle statistical methods. Epidemiologists should also look for consistency among the cases with the more acute events (for example, birth defects, neurotoxicity), and with evidence from biologic markers of exposure or disease.

In this discussion, we have briefly provided an orientation to cancer and its study from an environmental perspective. The complete coverage of carcinogenesis is a book unto itself, as is the concept of causation (Doll and Pieto 1981; Schottenfeld and Fraumeni 1982). However, this section is intended to acquaint the reader with some of the key concepts involved in carcinogenesis, and to provide a foundation for additional reading.

REPRODUCTIVE AND DEVELOPMENTAL EFFECTS

Historically, it was widely believed that the developing embryo was well secured from environmental insult within the protective and nurturing walls of the uterus. This notion has since been disparaged by a number of discoveries during the past several decades (Kalter and Warklany 1983). By the early 1930s, the adverse effects of intrauterine radiation on the developing fetus had been well documented (Goldstein and Murphy 1929). In 1941 the rubella virus was observed to cause congenital malformations among the offspring of mothers who had been exposed during pregnancy (Gregg 1941; South and Sever 1985). However, it wasn't until 20 years later that the thalidomide tragedy emphatically showed that the developing fetus was not free from potential insult from exogenous agents (McBride 1961; Lenz 1962).

The link between environmental contamination and adverse reproductive outcomes was first established in the 1950s with the discovery of a cluster of central nervous system malformations among newborn infants in Minamata Bay, Japan. The epidemic, which persisted for several years, was attributed to the ingestion by pregnant women of seafood contaminated with methyl mercury that had been dumped into the bay by local industries. More recently, incidents such as Seveso (Italy), Love Canal (New York), Alsea (Oregon), and Three Mile Island have sparked increased concern over the potential reproductive risks associated with exposure to environmental contaminants. Although many such episodes have been reported in recent years, in the majority of instances no clear connection between the exposure and alleged health outcomes could be established.

This is not to suggest that environmental factors are of little concern with regard to adverse reproductive outcomes. Indeed there are several reasons for continuing research directed at identifying and characterizing these risks. With the increasing number of new chemicals introduced each year into the environment, the number of agents that are likely to present a threat to reproductive health through their production, use, and disposal also increases. At the same time, an increasing number of reproductive-aged women are entering the workforce, with a growing number of these employed in hazardous industries and occupations. Because it is not presently possible to adequately test all such materials for potential health risks in humans, the number of individuals exposed to reproductive toxins and other hazardous agents is likely to increase in the future.

A second concern relates to the enhanced susceptibility of the developing fetus and young infant to environmental exposures. This is due primarily to the rapid rate of cellular differentiation and growth, and the development of complex structures such as the central nervous system. In addition to physiologic factors, the unique behavioral characteristics of infants and children may increase their risk for exposure to environmental toxicants such as lead (consumed in paint chips) or industrial effluent in neighborhood streams and ponds frequented by young children.

The effects of environmental insults on the newborn infant and child can result in a lifetime of disease, disability, and decreased productivity, thus placing an increased burden on already strained health resources. In addition to the clinically evident morbidity, an unknown number of adverse effects arise from chronic low-level exposures to certain agents. Much of this childhood morbidity may be preventable through early recognition and remediation of hazardous exposures.

Types of Adverse Reproductive Outcomes Associated With Environmental Exposures

Environmental agents may exert their effects at several points during the reproductive cycle, prior to or during conception and fertilization, prior to implantation, during embryogenesis (first trimester), or during the period of later fetal growth and development (second and third trimesters). As discussed in the following pages, the type of health outcome observed depends to a large extent on the timing of the exposure in relation to reproductive cycle.

The possible adverse reproductive effects that may potentially result from environmental exposures are varied, and range in severity from mild, transient effects (such as temporary amenorrhea) to highly lethal conditions (such as major chromosomal disorders). Table 9–3 presents some of the putative adverse outcomes that may be observed. The list is not a complete compilation of all possible effects; rather, it presents some of the more clinically significant outcomes, which could occur as a result of exposure to environmental factors. It should be noted that known ambient environmental factors probably play a relatively small role in the etiology of most of these conditions. For example, only about 5 to 8 percent of congenital malformations are known to be caused by environmental exposures such as drugs, chemicals, and ionizing radiation (Oakley 1986). In the vast

TABLE 9-3. Potential Adverse Reproductive Outcomes.

1. Delayed conception/infertility (sperm abnormalities, ovulatory or menstrual disorders).
2. Sexual dysfunction (decrease libido, impotence).
3. Spontaneous abortion (pregnancy loss prior to 28 weeks gestation).
4. Fetal mortality (fetal loss after 28 weeks gestation).
5. Neonatal mortality (infant deaths up to 28 days of age).
6. Low birth weight (preterm delivery, fetal growth retardation).
7. Single gene mutations (dominant lethal mutations, Mendelian disorders, other phenotypically manifest conditions).
8. Chromosomal abnormalities/aberrations (aneuploidy, nondisjunctive disorders, breaks, deletions).
9. Congenital malformations (single defects, multiple malformation syndromes/associations).
10. Developmental disabilities.
11. Childhood cancers.

majority (65 to 70 percent) of cases of congenital anomalies, the cause is unknown.

It is also important to recognize that many of these outcomes are interrelated, often in highly complex ways, and that these relationships may be influenced by the nature, timing, and intensity of exposure. For example, exposure to ionizing radiation during the period of organogenesis may result in fetal malformations, while exposures later in pregnancy have been linked to certain childhood malignancies. A low-level dose of radiation (e.g., < 5 rad) during the first trimester may produce no apparent adverse effects. At a higher dose there may be evidence of fetal anomalies, while at extremely high levels spontaneous abortion may result. Unlike carcinogens, most teratogenic agents (agents that cause congenital malformations) appear to exhibit a threshold of exposure below which no adverse effects are apparent.

Environmental Exposures Associated With Reproductive Disorders

A number of infectious (Rubella, cytomegalovirus), chemical agents (environmental pollutants, drugs), and physical agents (ionizing and nonionizing radiation, heat, noise) have been associated with deleterious reproductive effects in humans. Some of the major occupational and environmental agents known or suspected to be associated with adverse reproductive effects are presented in the Table 9–4.

TABLE 9-4. Environmental Agents Implicated in Adverse Reproductive Outcomes.

Exposure	Known/Suspected Effect
Anesthetic compounds	Infertility, spontaneous abortion, fetal malformations, low birth weight
Antineoplastics	Infertility, spontaneous abortion
Dibromochloropropane	Sperm abnormalities, infertility
Ionizing radiation	Infertility, microcephaly, chromosomal abnormalities, childhood malignancies
Lead	Infertility, spontaneous abortion, developmental disabilities
Manganese	Infertility
Organic mercury	Developmental disabilities, neurological abnormalities
Organic solvents	Congenital malformations, childhood malignancies
PCBs, PBBs	Fetal mortality, low birth weight, congenital abnormalities, developmental disabilities

Recognition and Evaluation of Environmental Reproductive Hazards

Since the thalidomide episode in the 1960s, two major approaches have been used in an attempt to identify potential embryotoxic or teratogenic agents: Animal testing and epidemiologic surveillance of human populations. To date, neither approach has proven to be of great value in providing an early warning of new reproductive hazards. The usefulness of *in vivo* testing has been limited to a large extent by the lack of a suitable animal model for humans, and the variation in susceptibility and response among different species/strains of test animals (Brent 1964; Fraser 1977). Despite these limitations, animal studies are frequently used by regulatory agencies as criteria for determining the potential teratogenicity of chemical compounds in humans (Brent 1986a). Laboratory studies may also be useful for elucidating the biologic mechanisms of action and pharmacodynamic properties of known human teratogens, and for generating leads for etiologic studies in humans (Brent 1986b).

Numerous congenital malformation registries and surveillance systems have been established throughout the world since the early 1960s. Although many of these programs were developed as early warning systems for detecting new human teratogens introduced into the environment, they have not proven to be particularly useful for this purpose (Kallen et al. 1984; Holtzman and Khoury 1986). Most teratogenic agents produce complex phenotypic manifestations such as malformation syndromes or associations that may be extremely difficult to detect with conventional surveillance techniques (Khoury et al. 1987). In addition, because the proportion of women initially exposed to a newly introduced teratogenic agent is often quite small relative to the size of the surveillance population, it may take several months or even years before an epidemic is detected. Once an alleged association is suspected, congenital malformation registries may be employed as databases for the conduct of etiologic epidemiologic studies.

Many of the environmental reproductive hazards recognized in the last few years were recognized by alert clinicians. However, these associations, which are frequently published as case reports, rarely provide sufficient evidence for causality. Well-controlled etiologic studies are required for assessment of a cause-effect association. Once a potential hazard has been identified, the decision whether to proceed with a more detailed investigation will depend on factors such as the biological plausibility of the association, the suspected magnitude of risk, the size of the exposed population, as well as the level of public concern surrounding the issue (MOD 1981).

NEUROLOGIC EFFECTS

Environmental exposures to neurotoxicants through occupational use, accidental spills during distribution or application, improper storage and disposal practices, and improper use practices pose significant public health risks. Neurotoxicants can be classified according to their activity at the anatomic site (Table 9-5).

Protective Mechanisms

The nervous system is thought to be protected from many environmental toxicants by the **blood-brain barrier (BBB).** The constituents of cerebrospinal fluid are not exactly similar to those of the extracellular fluid in other parts of the body. In fact, many large molecules will not be permitted to pass from the blood into the cerebrospinal fluid, even though these same substances may pass easily into the interstitial fluids of the body. The BBB exists in the parenchyma and the choroid plexus of the brain, except for an area of the brain that is particularly sensitive to changes in extracellular fluid components (that is, the hypothalamic area). The BBB is highly permeable to water, carbon dioxide, and oxygen, slightly permeable to electrolytes such as sodium, chloride, and potassium, and almost totally impermeable to metals such as arsenic, sulfur, and gold.

There are several hypotheses as to the protective nature of the BBB. One is that glial cells (for example, astrocytes, oligodendrocytes, and microglia) and astrocytic processes (called astroglia) found in the capillary endothelium function as barriers; where these cells are missing (for example, parts of the hypothalamus and the fourth ventricle) exogenous substances appear to have easier access to the neurons. An alternative hypothesis suggests that the extracellular membrane between the endothelial cells of the capillary and the glia and neurons may function to filter out foreign substances. One additional hypothesis suggests that characteristic properties of the brain

TABLE 9-5. Tissue-specific Damage by Neurotoxicants.

1. Myelin sheath, affecting the oligodendrocytes or Schwann cells.
2. Selected tissue of the central nervous system (CNS).
3. Neurons and astrocytes (of the gray matter) resulting from anoxia (absence or lack of oxygen).
4. Axons of peripheral neurons.
5. Synaptic junctions of the neuromuscular system.
6. Perikarya of the peripheral neurons.

Source: Norton (1980).

endothelial cells may provide a barrier (for example, the capillary endothelial cells are tightly spliced together by zonulae occludentes to prevent some large molecules from entering the brain tissue). Nonetheless, small molecules may still reach the neurons through the endothelial cells and glia.

It is also possible that pinocytotic vesicles, normally missing in endothelial tissue, may be found there as a result of some pathologic condition that would tend to increase the permeability of the BBB (for example, this condition is thought to occur in hypertension, ischemia, and post-irradiation) (Norton 1980). The brain also has less blood flow in the white matter compared to the gray. This difference in vascularization affects hypoxia, but the myelinated axons of the white matter, which require less oxygen than the gray matter, are less susceptible to toxic insult.

Neurons in some tissue (for example, dorsal root ganglia, and autonomic ganglia) of the peripheral nervous system (PNS) are susceptible to toxicants that slip through spaces between the epithelial cells. There are several other factors that might account for the selective sensitivity of some toxicants for certain anatomical areas of the CNS and the PNS that must yet be researched (for example, the damaging effects on neurons of amino acids; the high concentrations of acetylcholine, serotonin, etcetera, in the hypothalamus and other areas of the brain). However, it can be said that neurologic damage results from the selective penetration of some toxic chemicals through the protective barriers of the CNS and the PNS to specific anatomic sites (Table 9-6).

The most serious (and irreversible) damage occurs at the death of the neuron, since the differentiated cells cannot divide and be replaced. When neurons die, other cells with the same function (that is, fortunately, the nervous system is replete with redundancy of function) may be available to replace the dead cells, or other neurons may acquire the necessary function. If this repair mechanism fails due to the severity of the toxic insult, then loss of function will occur. Generally, after neurotoxic insult, some recovery of function occurs. When cell death is not involved, recovery occurs quickly (after the toxic agent is metabolized through the system) or biological regeneration has taken place. When neuronal death occurs, recovery may take longer or not occur at all. The more severe results from loss of function will be discussed in the following pages.

Physiology

Cell damage: Cellular damage from toxic chemicals occurs through (1) direct contact with the chemical or (2) a loss of oxygen to the cell. Loss of oxygen results in swelling of the cell and cytoplasmic organelles, dispersion of the rough endoplasmic reticulum (RER), and swelling of the nucleolus

TABLE 9-6. Damage to Selected Anatomic Areas of the Nervous System by Chemical.

	GRAY MATTER TYPE 1 ANOXIA			PROLONGED ANOXIA				WHITE MATTER TYPE 2				PERIPHERAL NEUROPATHY TYPES 3, 4, 5			LOCAL AREAS TYPE 6			
	Cortex IV	Cortex V	Hippocampus H1	Caudate Nucleus	Putamen	Hippocampus H2	Fascia Dentata	Globus Pallidus	Subthalamus	Internal Capsules	Corpus Callosum	Optic Chiasm	Schwann Cells	Sensory N. Thalamus	Anterior Horn Cells	Peripheral Axons	Hippocampus H3	Hypothalamus Ventral N.	Mammillary Body	Tegmentum
Acetylpyridine						+								+			+			
Acrylamide			+ +	+	+								+							
Azide	+ +		+	+	+				+											
Barbiturate																				
Carbon disulfide								+ +	+	+	+	+ +	+		+					+
Carbon monoxide	+							+												+
Cyanide								+												
DDT																				
Glutamate											+	+						+	+	
Gold thioglucose																				
Hexachlorophene											+	+	+ +		+ +					
Iminodipropionitrile																				
Isoniazid																				

Lead (inorganic)								
Adults	+	+						
Children						+	+	
Malononitrile	+	+				+	+	
Manganese	+	+				+	+	
Mercury (organic)	+				+		+	
Methyl bromide	+					+	+	
Nitrogen trichloride	+	+	+					
Pyrithiamine				+				
Triethyltin						+	+	+
Triothocresyl phosphate					+		+	
Vinca alkaloids								+ +

Source: Norton 1980.

(Figure 9-6). These effects are attended by a reduction in cytoplasmic pH, in activity of the oxidative enzyme systems, and in synthesis of protein and other cell constituents (Norton 1980). The neuron does not have much capability for anaerobic metabolism and a high metabolic rate. This simply means that in a situation where a neuronal cell is starved for air (anoxia) there is a tendency to rapidly metabolize a substance through the cell membrane. Generally, neuronal damage occurs rapidly, as blood flow is reduced and anoxia develops. Neurons, cells of the capillary endothelium, microglia, oligodendrocytes, and astrocytes are all highly sensitive to anoxia.

Anoxia can be divided into three classes.

1. Ischemic anoxia that results from a decrease in arterial blood flow that diminishes the flow of blood to the brain resulting in an accumulation of lactic acid, ammonia, and inorganic phosphate. Chemicals causing cardiac arrest will obviously result in loss of blood flow to the brain.
2. Cytotoxic anoxia occurs when the cellular metabolism is compromised through contact with metabolic inhibitors such as cyanide, azide, malononitrile, etcetera.
3. Anoxic anoxia occurs when there is a lack of oxygen with an adequate blood flow. For example, when the oxygen carrying capacity of blood is compromised, as is the case with the production of carboxyhemoglobin by carbon monoxide or methemoglobin by nitrites, the result is cellular damage, neuronal death. If the intoxication is sufficiently severe, coma and death of the individual will occur. Neuromuscular blocking agents may also create a situation where the bloods capacity to carry oxygen is reduced. This occurs when chemicals such as OP insecticides cause respiratory paralysis through interference with acetylcholine (the chemical transmitter at the neuromuscular junction).

Chemically related poisonings (intoxications) are considered to be a significant hazard to workers and others who are inadvertently exposed to toxic chemicals (for example, garden variety pesticides, solvents, and fertilizers). Organophosphorus and carbamate insecticides are felt to be particularly hazardous to workers and others who may become exposed (Popendorf and Leffingwell 1982; Davies et al. 1976).

Organophosphorus and carbamate pesticides are powerful inhibitors of plasma cholinesterase (PChE) and red blood cell cholinesterase (RBChE). For an excellent review of the physiology of cholinesterase (ChE) inhibition as a result of OP and carbamate exposure, read Durham and Hayes (1962) and Vandekar (1980). Briefly, the liver produces ChE, which is then found in the tissues of the PNS and the CNS, in the erythrocytes (RBChE) and in plasma (PChE) of circulating blood.

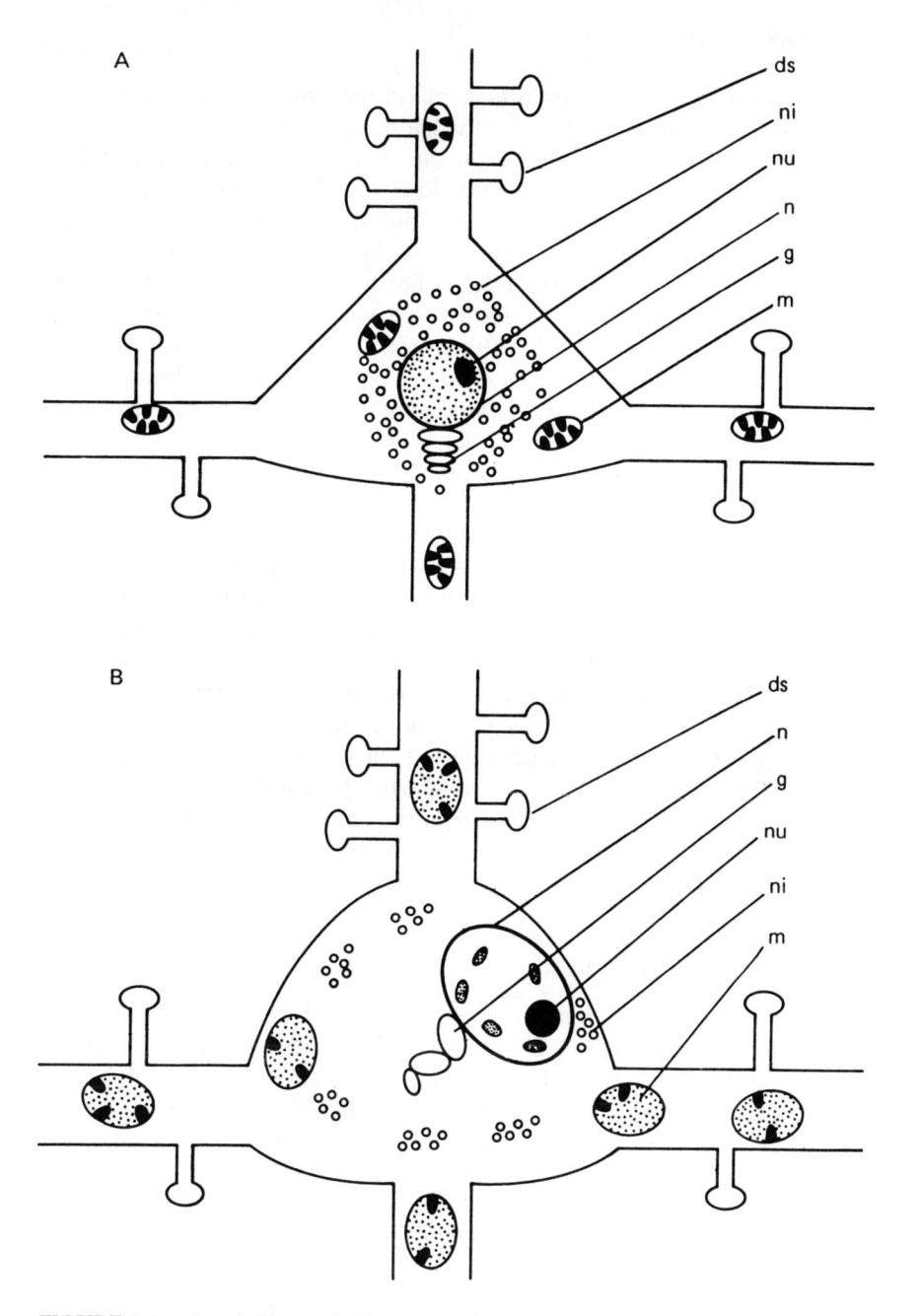

FIGURE 9-6. Anoxic Changes in Neurons. A. Normal Neuron: ds, Dendritic Spine; ni, Nissl Substance (RER); nu, Nucleolus; n, Nucleus; g, Golgi Substance; m, Mitochondrion. B. Anoxic Neuron: Swelling of Dendritic Spines, Mitochrondria, Nucleus, Nucleolus, and Golgi Substance; Clumping of Chromatin in the Nucleus (Pyknosis) and Dispersion of Nissl Substance (Chromatolysis). Lysosomes (Not Shown Also Swell). Source: By permission. Norton 1980.

Cholinesterase plays an important role in the transmission of nerve impulses by performing as a catalyst in the hydrolysis of a reversible acetic acid ester of choline. This process aids in the transmission of nerve impulses (acetylcholine) to choline and acetic acid at the cholinergic neuro-effector junctions, neuromuscular junctions, and in autonomic ganglia. (Guyton 1961; Davidson and Henry 1974).

Plasma cholinesterase can be distinguished from RBChE by the difference in substrate specificity between the two enzymes and by the use of ChE specific inhibitors (Witter 1963). Although PChE is not inhibited by excess levels of acetylcholine, RBChE is since it is bound to the membrane of the cell. The RBChE is thought to be the best indicator of acetylcholinesterase activity at the nerve synapse since it closely parallels the level of ChE in the CNS and PNS (Zavon 1965). However, PChE is possibly the most sensitive indicator of OP and carbamate exposure (Bogusz 1963; Rider et al. 1969).

Although OPs are responsible for an almost irreversible phosphorylation of RBChE that may require weeks of recovery, carbamylation of the enzyme is quickly reversible ($\leq$ 48 hours). Although the easy reversibility of carbamates suggests less toxicity, the effectiveness of cholinesterase measurement as a monitoring tool is reduced.

Plasma cholinesterase depression reflects recent and moderate exposure to OPs and carbamate pesticides. However, a significant reduction in both PBChE and PChE enzyme values suggests continuing exposure, or perhaps even one very large acute exposure. During monitoring activities, when RBChE and PChE values are found to to be 30 percent and 40 percent below a well-established personal baseline level, workers should be removed from contact with cholinesterase inhibiting chemicals.

Acute Neurological Effects

Symptomatology. Symptons of pesticide poisoning include eye and skin irritation, miosis, blurring vision, headache, anorexia, nausea, vomiting, increased sweating, increased salivation, diarrhea, abdominal pain, slight bradycardia, ataxia, muscle weakness and twitching, and generalized weakness of respiratory muscles. Central nervous system involvement is noted by giddiness; anxiety; restlessness; drowsiness; difficulty concentrating; poor recall; confusion; slurred speech; convulsions; coma with absence of reflexes; bursts of slow waves of elevated voltage in EEG, especially on overventilation; Cheyne-Stokes respirations; depression of respiratory and circulatory centers, with dyspnea cyanosis; and fall in blood pressure. In the latter stage of more severe poisonings, symptoms may progress from difficult and labored breathing to loss of muscle control, convulsions, and possibly death (Griffith et al. 1985).

Case History. A 57-year old black male reported dermal exposure to diuron and bromacil while applying the pesticide by ground-rig application. The worker said the spray drift struck his eyes and skin. The worker was treated by a physician for topical eye and skin irritation, and there were no lasting effects. The worker returned to work after treatment.

Generally, symptoms occur within 12 hours of exposure to the toxicant; however, delayed neurotoxicity is known to occur, days and months after exposure has taken place.

Case History. A 44-year old white male developed tetraplegia (paralysis of all four extremities) nine days after spraying his barn with an organophosphate insecticide, and expired due to respiratory failure four days later. Headache was the first symptom, occurring two days after exposure. He continued to work for several days, and then one week after exposure he became restless and developed an unsteadiness (weakness) in his right leg. The following day the right leg and right arm were paralyzed. Tetraplegia followed, with subsequent respiratory difficulties, and death in respiratory failure on the 13th day following exposure.

Case History. A 49-year old black male employed as an exterminator was exposed to several organophosphate insecticides. Shortly after spraying diazinon he began to experience vision problems, diplopia, lacrimation, excessive salivation, hyperhidrosis, muscle fasciculations, and cramps. He was also experiencing nausea, vomiting, abdominal cramps, and palpitations. The patient noted that during weekends and vacation periods he would be asymptomatic and would also feel "stronger." The patient developed a generalized myopathy and died in cardiac and respiratory failure approximately seven years after the initial symptoms occurred.

In the United States, not much is known about the actual number of occupationally related pesticide poisonings that occur annually. In 1976, Griffith et al., in a national sample of more than 8,000 general care hospitals, estimated that there were approximately 3,000 occupationally related pesticide poisonings that required hospital admission each year.

Many people believe that the number of poisonings is underestimated because workers are reluctant to report them for fear of losing their jobs, because poisonings are not properly diagnosed, and because there is no formal or mandated reporting requirement for physicians to report poisonings (except in California), as there is for certain infectious diseases. Other people believe that the problem is overstated, and that pesticide poisonings are not a serious national problem.

SUMMARY

This chapter has presented some baseline information regarding three of the more frequently discussed health effects associated with exposure to toxic materials via the ambient environment. These descriptions are simply primers in these three types of health consequences of environmental contamination. Further reading is urged.

One should recall that disease is a continuum, meaning that in a population with uneven levels of exposure (and with individual variability in susceptibility) a variety of health effects may occur from the same environmental exposure. Careful inspection of all health data—not just one type (for example the cancer incidence)—is a prudent course, with studies around a suspected or known environmental point source. If conditions or diseases are found that are not familiar to you, seek the counsel of physicians and researchers who specialize in this clinical manifestation. As with statistical guidance, advice concerning the evaluation of diverse health effects is also strongly encouraged.

Now consider that not all effects from exposure to hazardous materials are chronic or even life threatening. Some effects from toxic materials are evident in a short time (for example in days or weeks). These conditions may include dermatologic eruptions (for example rashes) as well as digestive or allergic reactions. Also, upper respiratory effects may occur, depending on the agent(s) and the route of exposure. When the Agency for Toxic Substances and Disease Registry surveys the members of their exposure registries each year, the survey contains questions on recurrent symptoms that the members have experienced as well as conditions for which they have sought medical attention. These measures are attempts to determine if, and to what extent, acute health effects may manifest themselves in settings of known hazardous exposure.

Effects such as those described in this section may be **reversible**—that is, if the exposure is stopped, the condition will disappear. With neurotoxicity, one considers the effects to be **nonreversible**—that is, when the exposure ceases, the condition will stabilize (with damage), but not continue to worsen. In the instance of cancer, the effects of a hazardous exposure are **irreversible**— that is, the deleterious course will continue even after the exposure is ceased. These categorizations of effects makes one pause to consider the potential for accruing health effects that may result from environmental disasters (for example, spills, fires, explosions).

ASSIGNMENTS

1. Describe an ecologic analysis that you might conduct (assume whatever database needs you require are available) to evaluate an environmental risk for a rare cancer of your choice. Include how you would define your cases and how you would develop a comparison experience to minimize confounding and secular trends.
2. Prepare a 50-word explanation for why a particular agent might cause liver

cancer among exposed workers, and mid-line defects (for example, cleft lip) among exposed fetuses *in utero*.
3. Discuss the types of exposure questions you would add to a special study nested within a National Exposure Registry if the acute effect observed was dermatitis, and lymphoma was the suspected cancer associated with the exposure.
4. Describe the acute symptoms that you would expect to find in a person who has been poisoned by an OP or carbamate insecticide.
5. Describe anoxic anoxia and how it may impact on a person who has been poisoned by a chemical compound, such as an OP insecticide.

REFERENCES

Allman, W. F. 1985. "We Have Nothing to Fear (But a Few Zillion Things)." *Science* 85:38–41.

Bailar, J. C., and Smith E. M. 1986. "Progress Against Cancer?" *New England Journal of Medicine* 314:1226–32.

Ballantyne, B. 1985. "Evaluation of Hazards from Mixtures of Chemicals in the Occupational Environment." *Journal of Occupational Medicine* 27(2): 85–94.

Bogusz, M. 1968. "Influence of Insecticides on the Activity of Some Enzymes Contained in Human Serum." *Clin Chem Acta* 19:367–69.

Brent, R. L. 1964. "Drug Testing in Animals for Teratogenic Effects: Thalidomide in the Pregnant Rat." *Journal of Pediatrics* 64:762–70.

Brent, R. L. 1986a. "The Complexities of Solving the Problem of Human Malformations." *Clinical Perinatology* 13:491–503.

Brent, R. L. 1986b. "Evaluating the Alleged Teratogenicity of Environmental Agents." *Clinical Perinatology* 13:609–13.

Cancer Facts and Figures. 1991. American Cancer Society, Atlanta, GA.

Cohn, B. A., Kaplan, G. A., and Cohen, R. D. 1988. "Did Early Detection and Treatment Contribute to the Decline in Ischemic Heart Disease Mortality? Prospective Evidence From the Alameda County Study." *American Journal of Epidemiology* 127:1143–54.

Davidson, I. and Henry, J. B. (eds) 1974. *Clinical Diagnosis.* Philadelphia: W. B. Saunders Co.

Davies, J. E., Shafik, M. T., Barquet, A., Morgade, A., Danauskus, J. K. 1976. *Residue Rev.* 62:45–57.

Davis, D. L., Hoel, D. Fox, J., and Lopez, A. 1990. "International Trends in Cancer Mortality in France, West Germany, Italy, Japan, England and Wales, and The USA." *Lancet* 336:474–81.

Dobson, A. J., Gibberd, R. W., Leeder, S. R., et al. 1988. "Ischemic Heart Disease in the Hunter Region of New South Wales, Australia, 1979–1985." *American Journal of Epidemiology* 128:106–15.

Doll, R and Peto, R. 1981. *The Causes of Cancer: Quantitative a Estimates of Avoidable Risks of Cancer in the United States Today.* New York: Oxford University Press.

Durham, W. F., and Hayes, W. J. 1962. "Organic Phosphorus Poisoning and Its Therapy." *Archives of Environmental Health* 5:21–43.

Fraser, F. C. 1977. "Relation of Animal Studies to the Problem in Man." In *Handbook of Teratology Vol 1*, ed. J. C. Wilson and F. C. Fraser. New York: Plenum Press.

Freudenburg, W. R. 1988. "Perceived Risk, Real Risk: Social Science and the Art of Probabilistic Risk Assessment." *Science* 242:44–9.

Goldstein, N. M., and Murphy, D. P. 1929. "Microcephalic Idiocy Following Radium Therapy for Uterine Cancer During Pregnancy." *American Journal of Obstetrics and Gynecology*:189–95.

Gregg, N. A. 1941. "Congenital Cataract Following German Measles in the Mother." *Trans. Opthalmol. Soc. Aust.* 3:35–46.

Griffith, J., and Duncan, R. C. 1983. "An Assessment of Fieldworker Occupational Exposure To Pesticides in The Florida Citrus Industry." In: *National Monitoring Study: Citrus. Vol. III*. Miami, Florida: University of Miami Press.

Griffith, J., Duncan, R. C., Konefal, J. 1985. "Pesticide Poisonings Reported by Florida Citrus Fieldworkers." *Environ Sci Health* B20(6):701–27.

Griffith, J., Vandermer, H., Blondell, J. 1976. *National Study of Hospital Admitted Pesticide Poisonings*. Epidemiologic Studies Program, Office of Pesticide Programs, U.S. EPA, Washington, D.C.

Guyton, A. C. 1961. *Textbook of Medical Physiology*. Philadelphia: W. B. Saunders Co. pp. 237–43.

Hall, J. M., Ming, K. L., Neuman, B., Marrow, J. E., Anderson, L. A., Huey, B., and King, M. 1990. "Linkage of Early-Onset Familial Breast Cancer to a Chromosome 17q21." *Science* 250:1684–89.

Holtzman, N. A., and Khoury, M. J. 1986. "Monitoring for Congenital Malformations." *Ann. Rev. Pub. Health* 7:237–66.

Horowitz, R. I., and Feinstein, A. R. 1978. "Alternative Analytic Methods For Case-Control Studies of Estrogens and Endometrial." *New England Journal of Medicine.* 299(20):1089–94.

Howe, H. L. 1988. "A Comparison of Actual and Perceived Residential Proximity to Toxic Waste Sites." *Archives of Environmental Health* 43:415–19.

Kallen, B. Hay, S., and Klingberg, M. 1984. "Birth defects monitoring systems. Accomplishments and Goals." In *Issues and Reviews in Teratology. Vol. 2*, ed. H. Kalter, 1–22. New York: Plenum Press.

Kalter, H., and Warklany, J. 1983. "Congenital Malformations. Etiologic Factors and Their Role in Prevention (Part 1)." *New England Journal of Medicine* 308:424–31.

Kaplan, G. A., Cohn, B. A., Cohen, R. D., et al. 1988. "The Decline in Ischemic Heart Disease Mortality: Prospective Evidence From the Alameda County Study." *American Journal of Epidemiology* 127:1131–42.

Khoury, M. J., Adams, M. M., Rhodes, P., and Erickson, J. D. 1987. "Monitoring for Multiple Malformations in the Detection of Epidemics of Birth Defects." *Teratology* 36:345–53.

Kinzler, K. W., Nilbert, M. C., Vogelstein, B., et al. 1991. "Identification of a Gene Located at Chromosome 5q21 that is Mutated in Colorectal Cancers." *Science* 251:1366–70.

Lenz, W. 1962. "Thalidomide and Congenital Abnormalities." *Lancet* 1:45.

Marx, J. L. 1989. "How Cancer Cells Spread in the Body." *Science* 244:147–8.

McBride, W. G. 1961. "Thalidomide and Congenital Abnormalities." *Lancet* 2:1358.
Mollgavkar, S. H. 1983. "Model for Human Carcinogenesis: Action of Environmental Agents." *Environmental Health Perspectives* 50: 285–91.
MOD. March of Dimes. 1981. "Report of Panel II. In *Guidelines for Studies of Human Populations Exposed to Mutagenic and Reproductive Hazards.*" A. Bloom ed. Proceedings of conference held January 26-27, 1981. Washington, D.C., White Plains: March of Dimes.
Murphy, S. D. 1980. "Assessment of the Potential for Toxic Interactions Among Environmental Pollutants." In *The Principles and Methods in Modern Toxicology*, ed. C. L. Galli, S. D. Murphy, R. Paoletti, pp. 277–94. Elsevier/North-Holland Biomedical Press.
Norton, Stata. 1980. "Toxic Responses To The Central Nervous System." In *Casarett and Doull's Toxicology*, eds. J. Doull, C. D. Klaassen, and M. O. Amdur. New York: Macmillan Publishing Co., Inc.
Oakley, G. P. 1986. "Frequency of Human Congenital Malformations." *Perinatology* 31:297–554.
Popendorf, W. J. and Leffingwell, J. T. 1982. *Residue Reviews* 82:125–201.
Rider, J. A., Moeller, H. C., Puletti, E. J., Swader, J. I. 1969. "Toxicity of Parathion, Systox, Octamethyl-Pryophosphoramide, and Methylparathion in Man." *Toxicology and Applied Pharmacology* 14:603–11.
Ruckelshaus, W. D. 1984. "Risk In A Free Society." *Risk Analysis* (4):157–62.
Schottenfeld D., and Fraumeni J. 1982. *Cancer Epidemiology and Prevention*. Philadelphia: W. B. Saunders Co.
Schechter, M. T., Spitzer, W. O., Hutcheon, M. E. et al. 1989. "Cancer Downwind from Sour Gas Refineries: the Perception and the Reality of an Epidemic." *Environmental Health Perspectives* 79:283–90.
SEER. 1973-1987. *Cancer Statistics Review*. National Cancer Institute. DCPSC, NIH Pub. No. 90-2789:1989.
Shapiro, S., Levine, H. S., and Agramowicz, M. 1971. "Factors Associated With Early and Late Fetal Loss." *Advances in Planned Parenthood*. 6:45.
Shimkin, M. B. 1977. *Contrary to Nature*. National Cancer Institute, NIH Pub. No. 760-720.
Shubik, P. Progression and Promotion. 1984. *Journal of the National Cancer Institute* 73(5):1005–11.
South, M. A., and Sever, J. L. 1985. "Teratogen Update: The Congenital Rubella Syndrome." *Teratology* 31:297–307.
Vandekar, M. 1980. "Minimizing Occupational Exposure to Pesticides: Cholinesterase Determination and Organophosphorus Poisoning." *Residue Reviews* 75:67–78.
Weinstein, I. B. 1991. "Mitogenesis is Only One Factor in a Carcinogenesis." *Science* 251:387–88.
Williams, O. D., Rywik, S., Sznajd, J., et al. 1989. "Poland and US Collaborative Study on Cardiovascular Epidemiology." *American Journal of Epidemiology* 130:457–67.
Witter, R. F. 1963. "Measurements of Blood Cholinesterase." *Archives of Environmental Health* 6:537.
Zavon, M. R. 1965. "Blood Cholinesterase Levels in Organic Phosphate Intoxication." *JAMA* 192:51–21.

10
Risk Assessment

Jack Griffith, Tim Aldrich, and Wanzer Drane

OBJECTIVES

This chapter will:

1. Describe the components of risk assessment.
2. Describe the benefits and problems associated with the use of epidemiologic data in the risk assessment process.
3. Describe the assumptions made in calculating unit risk.
4. Demonstrate the calculation of a sample risk assessment based on animal data alone.
5. Demonstrate the calculation of a sample risk assessment based on epidemiologic data alone.
6. Describe the interpretation of risk assessment findings (that is, basic products of risk assessment).

PUBLIC POLICY

The public's concern over environmentally related health effects resulting from involuntary exposure to toxic pollutants has shaped the evolution of public policy in the United States. As citizens, and as individuals, we are all concerned that our health, and the health of our children, may be compromised or endangered by exposure to a myriad of toxic chemicals.

Unfortunately, the ability of health professionals to adequately respond to this concern has too often been limited by inadequate funding and limited technology. Media attention to environmental "episodes," while elevating the level of awareness on the part of the public regarding potential risks, has also been responsible for contributing to an inappropriate concern (in some cases) for the safety of exposed persons.

The insistence by the public that it be protected from environmental problems, and that the environment be protected as well, is of concern to public health and regulatory authorities since regulation often entails economic sacrifice on the part of some segment of the public (Bender et al. 1990). This concern is fueled by a dearth of health professionals who have the training and depth of experience required to successfully and objectively estimate human health risks in a population. This chapter is designed to aid health professionals and others to understand the technology involved in establishing risk estimates related to environmental pollutants.

Risk Assessment Defined

Risk assessment is the characterization of potential adverse health effects of human exposures to environmental hazards. The risk assessment process includes:

1. A description of the hazard based on an evaluation of available epidemiologic, toxicologic, clinical, and environmental research.
2. Extrapolation from existing data to predict the type of health risks that an exposure may pose.
3. An estimate of the magnitude of these potential health risks to humans under given exposure conditions.
4. An estimation of the number and characteristics of exposed persons, at various intensities and durations of exposure.
5. An estimation of the extent of the overall public health problem brought about by public exposure to a particular environmental hazard.

To complete the risk assessment process it is also necessary to characterize all of the uncertainties inherent in attempting to infer risk. Shown in Figure 10-1 is a diagram that depicts the elements of the risk assessment process. The elements of a process form a paradigm or model. The sequence of relationships within a process are implicit in the structure of the paradigm.

It is implicit in the process of risk assessment that risk can be quantified. Risk assessment usually begins with cellular organisms (*in vitro* analyses), progressing to laboratory animals (*in vivo* analyses), and on to human data via epidemiologic studies. Although epidemiologic data are considered to be the most convincing evidence of human risk in the regulatory process, risk assessments are usually performed with the first two levels of data.

Several factors are responsible for the lack of epidemiologic data in risk assessment:

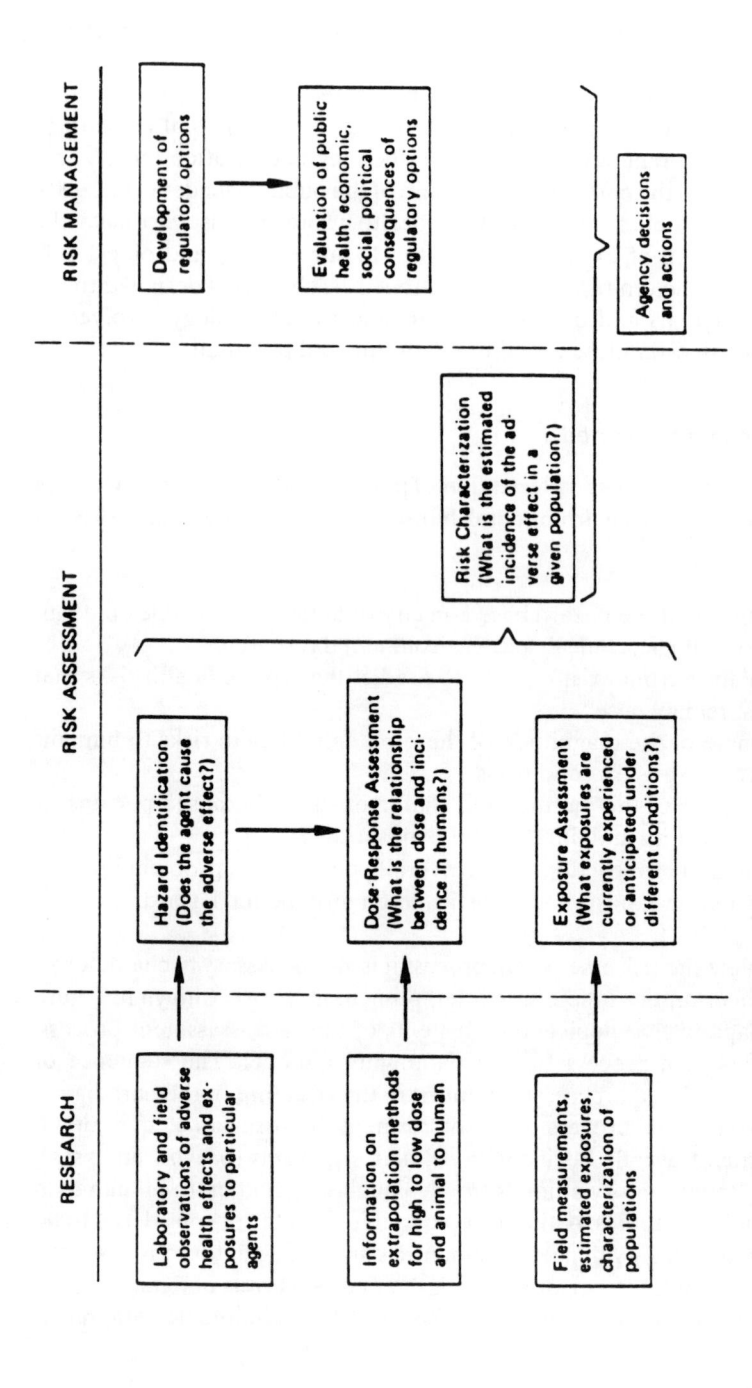

FIGURE 10-1. Major Elements of Risk Assessment and Risk Management. Reproduced with permission from National Research Council 1983.

1. Human data are difficult and expensive to generate.
2. In chronic disease situations one often finds very low risks (for example, twofold increased incidence).
3. The number of exposed persons is small.
4. The latency period between exposure and disease is often many years, and exposure characterization is difficult and often impossible.
5. Humans are often exposed to complex mixtures of chemical and physical agents.

Although more than 65,000 chemicals are presently in use, only a very few (< 1,000) have been reviewed epidemiologically. From the standpoint of risk assessment, it is also much easier to manipulate doses in a laboratory setting and to evaluate health effects (for example, tumors formed) with experimental systems than with free living humans.

The four components of risk assessment are hazard identification, dose-response estimation, exposure assessment, and risk characterization. **Hazard identification** is the process used to determine whether exposure to an agent may be responsible for an increase in specific health effects (for example, cancer, reproductive effects, neurological disorders).

The key to hazard identification is characterizing causation: Does the substance cause the disease? Of course, it often happens that no human data are available to make this determination, and it becomes necessary to use data from laboratory animal bioassays, or other tests (for example, *in vitro* mutagenic assays, or molecular structure comparisons). Unfortunately, extrapolation from animal to human models is not an exact science. In fact, not everyone believes that extrapolation from one animal system to another is feasible, much less from one animal species to another (for example, extrapolating from mice to man). However, in the regulatory process, positive answers to questions of disease risk in animal models are assumed to mean that the same agent may pose the same health risk to humans.

Dose-response estimation attempts to demonstrate that increasing disease risk is associated with increasing exposure. In the risk assessment paradigm, dose-response is the process of characterizing the relationship between the dose received and the incidence of adverse health effects in exposed populations. Dose-response then is the estimation of the incidence of the event expressed as a function of human exposure to the particular agent. The dose-response curves for mortality related to exposure to two chemicals, A and B, are shown in Figure 10-2. Although the LD_{50} is the same for both chemicals (8 mg/kg), the slopes of the response curves are very different. At one half of the LD_{50} of the chemicals (4 mg/kg) less than 1 percent of the animals exposed to chemical B would die, but 20 percent of the animals given chemical A would die (Klaassen and Doull 1980). The LD_{50} (amount of agent

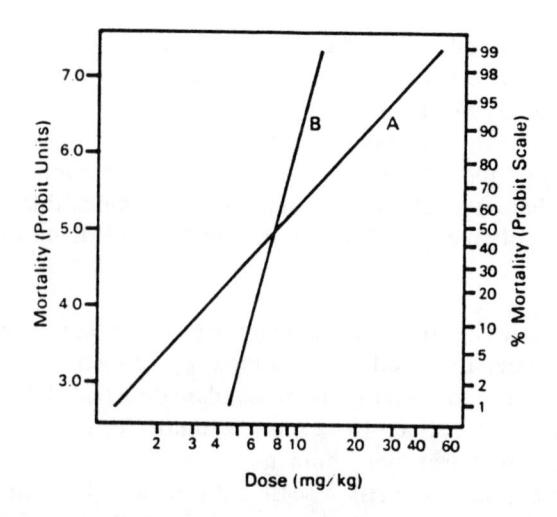

FIGURE 10-2. Diagram of Dose-Response Relationships. Chemical B Has a Much Steeper Dose Response Than Chemical A. Source: Klaassen and Doull, MacMillan Publishing Co., Inc.

required to produce death in 50 percent of the dosed animals) for a number of chemicals is shown in Table 10-1.

Efforts to determine whether a dose-response relationship exists involves consideration of the intensity and pattern of exposure (that is, schedule), the duration of exposure, as well as other factors that might play a role in an individual's response to dose. In Figure 10-3, we see a dose-response relation-

TABLE 10-1. Acute LD_{50}s for Selected Chemical Compounds.

Chemical	LD_{50} MG/KG
Chlordane	457–590
Chlorobenzilate	960
Paraguat	4
DDT	100
Ethyl alcohol	10,000
Sodium chloride	4,000
Ferrous sulfate	1,500
Phenobarbital sodium	150
Nicotine	1
Dioxin (TCDD)	0.001
Botulinus toxin	0.00001

Source: Abstracted from Klaassen and Doull (1980).

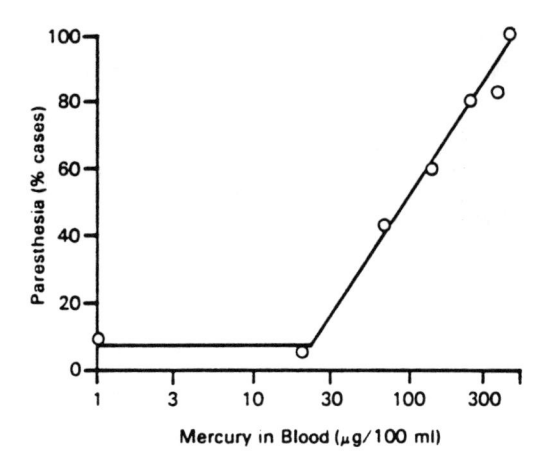

FIGURE 10-3. Dose-Response Relationship for Methyl Mercury, With the Concentration of Mercury in the Blood as the Dose, and Paresthesia (an Abnormal Sensation Such as Burning or Prickling of the Skin) as the Response. Source: By permission. Hammond and Beliles 1980.

ship between mercury in the blood and paresthesia (prickling or burning sensation). Dose-response relationships were developed from data gathered as a result of a large methyl mercury poisoning episode in Iraq. The concentration of mercury in the blood is an adequate expression of dose under long-term exposures. Clearly, as blood mercury levels increase, so does the percentage of paresthesia cases among the exposed population. Figure 10-3 also depicts a dose threshold, a level below which there is little or no adverse effect.

Exposure assessment is the process of measuring or estimating the intensity, frequency, and duration of human exposures to an environmental agent or of estimating exposures emanating from the release of chemicals into the environment. Exposure assessment describes the magnitude, duration, and route of exposure; characterizes the exposed populations by age, sex, race, and size; and addresses the uncertainties in all the estimates described. Exposure assessment may be used to predict the feasibility of regulatory control options and the effects of control technologies on reducing exposure. For example, in Figure 10-4, we see a comparison between lead used in gasoline and blood lead levels in preschool age children from the National Health and Nutrition Examination Survey (gathered from 1976–1980). Clearly, from 1976 through 1980, there was a significant reduction in lead used in gasoline and a corresponding decrease in blood lead levels of the children. It appears, from a cursory examination of Figure 10-4, that the regulatory decision by the U.S. Environmental Protection Agency to require lead free gasoline was extremely efficacious.

218   Environmental Epidemiology

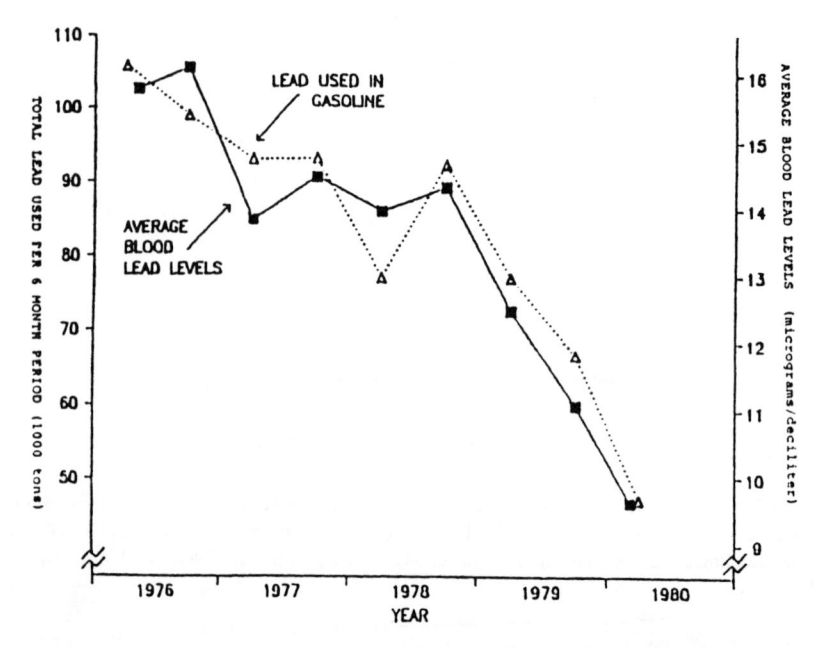

FIGURE 10-4.   Lead Used in Gasoline Production and Average NHANES II Blood Lead Levels, February 1976 to February 1980. Source: With permission. Goldsmith 1986.

Risk characterization is the process of estimating the incidence of an adverse health effect that may be expected from exposure to a specific environmental agent. Risk characterization uses the components described in the section on exposure assessment to determine how much of an agent may reach human populations and whether those persons may have more or less risk, based on their exposure characteristics (that is, continuous versus occasional exposure; children versus adults; air pollution versus water pollution, etcetera).

Risk characterization is performed by combining this exposure information with dose-response assessments (that is, what effects may result from the specific levels of exposure). A dose-response curve that gives equal weight to the dose rate and the duration of exposure is shown in Figure 10-5. The data reflect asbestos dust exposure over a continuum, and the lung cancer mortality of asbestos workers. It is clear from Figure 10-5 that as exposure to asbestos dust increases over time, lung cancer mortality also increases. Data like these are fairly easily generated in epidemiologic studies. However, to be very useful the data should provide information on duration of exposure, and the time when exposure began.

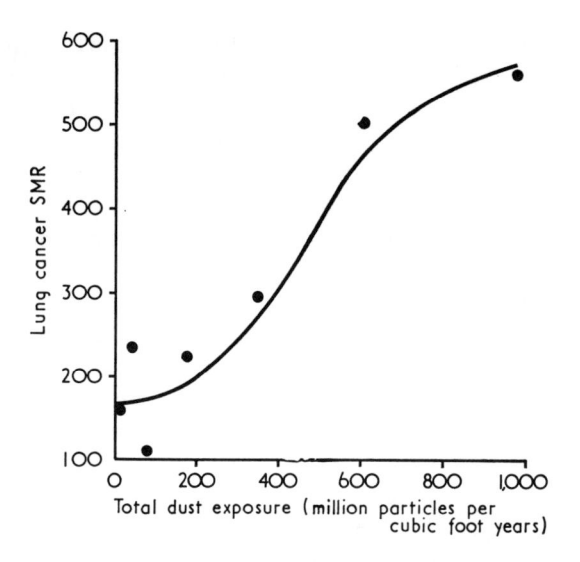

FIGURE 10-5. Total Asbestos Dust Exposure and Lung Cancer Mortality. Source: By permission from Enterline et al. 1973.

THE RISK ASSESSMENT PROCESS

The risk assessment process may vary somewhat, depending on the media or pollutant. For the purposes of this discussion, we will follow a commercially available pesticide through the entire process. Cholorobenzilate is a carbinol acaricide used primarily as an emulsifiable concentrate on oranges and grapefruit, and is sold under the trade names of Akar®, Folbex®, Acaraben®, Benzilan®, and Kop-Mite®. In the late 1960s, the Mrak Commission (an advisory committee to the Secretary of Health, Education and Welfare, now the Department of Health and Human Services) recommended that human exposure to chlorobenzilate be minimized and that use of this pesticide be **restricted** to those purposes for which there are benefits to human health that outweigh the potential hazard of carcinogenicity. The EPA considered the occupational use of chlorobenzilate to pose the greatest risk to human health (applicators, fruit pickers) (EPA 1979a), and subsequently issued a notice of Rebuttable Presumption Against Registration (RPAR) of pesticide products containing chlorobenzilate (41 *Federal Register* 21517, May 26, 1976).

The RPAR notice in the *Federal Register* signals completion of the process of hazard identification, and invites rebuttal of the decision by the EPA to discontinue registration of the pesticide. Parties wishing to continue to register this pesticide for manufacture and sale will then provide the EPA with specified health and safety data, as well as economic benefits associated with the use of this particular chemical. The RPAR process terminates when the EPA decides to either register, suspend, or cancel the registration of the product, or until the case has been adjudicated and a legal judgment rendered. If the rebuttal by the registrant is successful, then the product will be registered or reregistered.

After extensive hearings with the Science Advisory Panel (a group of scientists from academia and industry that advises the EPA on pesticide related matters) and interested parties (that is, U.S. Department of Agriculture, and industry representatives), the EPA proposed the cancellation and denial of registration of chlorobenzilate products for uses other than citrus (the primary citrus states are Florida, Texas, California, and Arizona). Registration of chlorobenzilate products for citrus use in those states was also to be canceled or denied unless registrants or applicants classified chlorobenzilate products for **restricted use.**

Restricted use means that the pesticide may only be applied by certified applicators (persons licensed to apply restricted use pesticides) or persons under their supervision. Applicants were directed by the EPA to submit "identified" exposure data to EPA within 18 months. The registrants and applicants were also directed to modify the label and to provide the following precautionary statement (EPA 1979b):

1. Take special care to avoid getting chlorobenzilate on eyes, skin or clothing.
2. Avoid breathing vapors or spray mist.
3. In case of contact with skin, wash as soon as possible with soap and water.
4. If chlorobenzilate gets on clothing, remove contaminated clothing, and wash parts of body with soap and water. If the extent of contamination is unknown, bathe entire body thoroughly, and change to clean clothing.
5. Wash hands with soap and water each time, before eating, drinking or smoking.
6. At the end of the work day, bathe entire body with soap and plenty of water.
7. Wear clean clothes each day, and launder before using.

The label was also required to recommend the following protective clothing be used during application:

1. One piece overalls which have long sleeves and long pants constructed of finely woven fabric as specified in the USDA/EPA Guide for Commercial Applicators.
2. Wide brimmed hat.
3. Heavy-duty fabric work gloves.

4. Any article of clothing worn while applying chlorobenzilate must be cleaned before reusing. Clothing which has been drenched or has otherwise absorbed concentrated pesticide must be buried or burned.
5. Use a facepiece respirator of the type approved for pesticide spray applications by the National Institute of Occupational Safety and Health.
6. Heavy-duty rubber or neoprene gloves and apron must be worn during loading, unloading and equipment clean-up.
7. Instead of clothing and equipment specified above, the applicator could use an enclosed tractor cab which provides a filtered air supply. Aerial application could be conducted without the specified clothing and equipment.

In attempting to develop an adequate level of protection for ground applicators, the EPA revised exposure estimates based on "exposure to forearms, hands, and face (15.8 percent) of the total body surface." It was the EPA's contention that "covering the forearms and hands would reduce dermal exposure from between 12 and 40 mg/day to between 4 and 10 mg/day." Facepiece respirators would effectively eliminate exposure by inhalation, estimated at 1 mg/day, and further reduce dermal exposure to the face by 1–3 mg/day. Thus, the total exposure could be reduced to between 2 and 6 mg/day (EPA 1979a). The EPA also estimated the maximum lifetime probability of tumor formation to be between 65 to 190 cancer cases in one million workers, based on a daily dose of 2–6 mg/day (40 days/year for 40 years).

How successful was the EPA in predicting exposure and risk? Griffith and Duncan (1980) monitored workers (applicators, mixers, loaders, and pickers) in Texas and Florida (Table 10-2), and found chlorobenzilate residue values in their urine to be 1.19 mg/day for the Florida workers, and 4.0 mg/day for the Texas workers. Florida spray season workers were well within the limits proposed as safe by the EPA. When both groups were combined the value (3.44 mg/day) was within the safety range proposed by the EPA. Based on the levels of chlorobenzilate residues in the urine of the Florida and Texas workers, tumor risk was projected to range from 22 to 202 cancer cases in one million workers (based on a daily dose of 3.44 mg/day (40 days/year for 40 years). The EPA had projected the risk to be 65 to 190 cases per million workers (based on a daily dose between 2 and 6 mg/day). Clearly, the risk estimate prepared by the EPA was remarkably close to estimates based on actual field monitoring data.

In another scenario, in attempting to regulate the discharge of toxic waste into the ambient air, the EPA identifies an environmental point source (for example, industry smoke stack, or a hazardous waste disposal site), and develops a "site profile." The site profile contains operational details about the facility, its location, and the population living near the

TABLE 10-2. Observed and estimated Chlorobenzilate Levels Among Citrus Workers in Florida and Texas.

		Urinary Residues (ppm)	Estimated Exposure (mg/day)	Estimated Dose (mg/day)
Workers				
Florida	mean	0.304	11.9	1.19
(n = 15)	std. error	0.107	5.2	0.52
	95% C.I.	0.00–0.66	1.0–22.8	0.10–2.28
Texas	mean	2.67	40.0	4.0
(n = 61)	std. error	1.27	17.5	1.75
	95% C.I.	0.18–4.63	5.6–74.4	0.56–7.44
Combined	mean	2.20	34.4	3.44
(n = 76)	std. error	1.03	14.2	1.42
	95% C.I.	0.18–4.22	6.7–62.2	0.67–6.22
RPAR Position Document 3			120–440	12–40
RPAR Position Document 4			20–60	2–6

Calculation: Urine volume was corrected by dividing the observed osmolality into an assumed "standard" osmolality of 800 mosm. Then a daily volume was calculated assuming an average urine output of 1,100 ml/day. Finally, the total estimated urinary output of chlorobenzilate in micrograms (μ) was calculated by multiplying the observed ppm (μg/ml) by the computed volume. The estimated dose level (amount deposited on the skin and available for absorption) was then calculated by using the recovery rate for dermal exposure to chlorobenzilate reported by Brady et al., of 5.51 percent. Finally, the actual dose was calculated as 10 percent of the estimated exposure according to Feldman and Maibach as noted in the EPA chlorobenzilate PD 3 (p. 28).

For example:

urinary residue = 0.14 ppm,
osmolality = 516 milliosmoles
then,

est. exposure $\frac{0.14 \, \mu g/ml \times 10^{-3} \, mg/g \times 800 \, mosm}{516 \, mosm \times 0.0515} \times 1,100 \, ml/day = 4.64 \, mg/day$,

and,
est. dose = 0.464 mg/day

site. In Figure 6-2, although there are many exposure pathways identified (arrows), rarely does on-site exposure monitoring take place. Usually, airborne, or water transport models are used to estimate the expected exposures.

The next step in the process is a risk characterization (Figure 10-1) of the actual risk posed by the hazard (for example, use of the pesticide, or emissions from the identified point source). As shown in Figure 10-1, the objective of risk assessment is to assess the intensity of the danger these exposures present to humans or to the environment (that is, the magnitude of the potential hazard).

At this point in the risk assessment process the risk characterization is performed by evaluating the magnitude of the exposure and the toxicity of the agent(s) (that is, exposure level at this particular site, and toxicity of the chemical pollutants). Usually, this assessment is based on experimental animal data, although human data **may** be used if available. **This then, is the setting for conventional risk assessment!**

When undertaking a risk assessment, you must begin by reviewing all information relevant to the particular exposure, including the toxicological, medical, and epidemiologic literature. This process will not be elaborated on at this time, except to say that all relevant information must be acquired on the subject exposure, or agent for which the risk assessment is to be performed. This process must not be minimized and several excellent documents are recommended for reading: Please refer to the "Recommended Reading" section. Questions to be answered **before** the risk assessment begins include:

1. What are the health effects associated with this exposure?
2. What are the routes of exposure for this chemical, or from this site?
3. Are particularly susceptible persons likely to be exposed?
4. Who will be exposed (animals, humans, workers, very susceptible individuals, such as children, pregnant women, older persons, etcetera)?
5. How many people will be exposed (is the setting rural, populous, industrial, residential, etcetera)?
6. What is the magnitude, duration and timing of exposure? This includes inspection of the probable pathways of exposure (air, water, soil, etcetera).
7. Based on animal studies is there a threshold (that is, an exposure level considered to be safe) level of exposure?
8. Based on the animal literature, should a dose-response relationship be expected?
9. Is the available animal exposure data suitable for extrapolation to man?

When these questions are answered, and other available data (for example, political, social, and economic impacts; engineering controls) are factored into the process, the risk assessment is ready to be calculated.

Assumptions Involved With Risk Assessment

To elaborate on the assumptions usually involved in answering these questions: Much of the available information on either a specific agent or a mixture of agents will be in the form of toxicology studies that use plant or animal test systems. From the early years of studying ionizing radiation, there are extensive mathematical models based on body weight and surface area for extrapolating animal doses to man. In general, the extrapolation is from the animal to that of a **standard** 70 kilogram (kg) man. The model should be adjusted for women, children, etcetera. However, this adjustment is almost never made, the risk assessor choosing to use the **generic human** for consistency across risk assessments of other compounds.

For many agents there is an accepted exposure level below which no adverse effects are observed. **This safe level is termed a threshold** and is an important consideration in regulatory decisions (note Figure 10.3). Often, in the interest of safety, an assumption is made that there is no threshold for a candidate substance or mixture. This assumption also holds for the suspected etiologic effect being used with the risk analysis. Generally, cancer is the health effect of concern with environmental risk assessments. The proposition for risk assessment with no threshold is based on the idea that a single cellular transformation (initiation) by this chemical is all that is required for cancer to occur.

A major assumption is involved with the issue of the dose-response relationship for the exposure of concern. In the following section, the issue of dose-response relationships is dealt with in greater detail. At this point, it is sufficient to note that for most risk assessment performed for a regulatory purpose, there is the assumption that the relationship between exposure and effect is linear: No threshold is considered and no differences in transport, metabolism, or effect are accepted for very low or very high exposures.

The process involved for establishing the unit risk (that is, lifetime exposure, worst-case scenarios, maximum permitted exposure over a lifetime, etcetera) in epidemiologic studies deserves further comment. In epidemiologic studies, we are concerned about place and time (of exposure) as they relate to disease (person). As shown in the previous example on chlorobenzilate, the EPA established a tumor risk for an exposed population based on **40 days of exposure over a 40 year period.** This is the usual method for calculating the unit of dose over time (i.e., the **cumulative dose** is developed by adding each exposure (dose) over time).

Measuring the cumulative dose is appropriate when one is attempting to measure xenobiotic agents that insult the host, perhaps at the target organ site, produce an effect (molecular change) and then disappear. However, if the toxic agent tends to remain in the tissues over an extended period of time, an estimate of cumulative dose is not sufficient. For example, lead residues remain in human tissues for years and continue to cause serious health related problems. It would be well to know, in predicting risk, what fraction of the tissue burden remains, as well as the amount of additional lead added to the system (that is, **the integrated dose**). To do this, we must understand the half-life (retention in the body) of the agent in the body, and its effects on biological functions, such as metabolism, storage, and excretion.

Four specific quantitative issues are presented here.

Extrapolation Methods. When a decision is made about human risk based on animal data, the issue of extrapolating from animal to man arises. Extrapolation is highly controversial since animal experiments are generally conducted at very high dose levels (posing the problem of extrapolating from high dose to low dose across species). The Lethal Dose$_{50}$ (a dosage that will kill 50 percent of the animals) is commonly known as the LD$_{50}$. For most toxicology experiments, animals are dosed just below their LD$_{50}$ to assure that there will be a biologic response (for example, tumors will appear, reproductive effects will take place). Such exceptionally high doses not only raise the possibility of potentially saturating detoxification pathways, they are greatly dissimilar to the low-level human exposures generally observed in the ambient environment.

When testing an agent for carcinogenicity, an experiment is set up with an exposed group and a control group of animals of the same species, age, sex. Test animals may be composed of one or two exposure levels. Usually, the high exposure level will be based on the ED$_{50}$ (over the years, as many, many chemicals have been tested for carcinogenicity, conventional wisdom has designated as the Effective Dose$_{50}$ (ED$_{50}$), an exposure producing tumors in 50 percent of the experimental animals). Generally, a bioassay of this type will have about 50 animals in each test group. The null hypothesis is that each group will have the same number of cancers; the alternative hypothesis is that the exposed animals will have more cancers (that is, this is a one-tailed statistical test). From conventional screening methodology, if the null hypothesis is not accepted, then the test results are considered "positive." With a cancer prevalence of 10 percent in the control animals (i.e., five cancers produced in a cohort of 50), the probability of a false positive (finding an increased risk when there is none) can be set at 0.05.

However, consider the impact to the study if only one cancer is actually caused by the exposure (and only one cancer [2 percent] is expected in the control group). In this situation, the false positive rate declines to 0.02. What do you think would be the impact for these hypothetical situations in terms of the false negative rate (saying there is no difference in cancer rates because of the exposure, when in fact there is an increased rate)?

To consider the impact on false negative rates, let's use the experimental configuration shown above (that is, 50 animals in each group, 10 percent and 2 percent prevalence respectively). For the 10 percent prevalence, the false negative rate is 0.00! But for the 2 percent prevalence, the chances of making a false negative error is 68 percent! Now, one begins to appreciate the procedures used for carcinogen screening employed by the National Cancer Institute. The public's health must be protected, even to the potential detri-

226   Environmental Epidemiology

ment of specific agents. This means that false negatives are much less acceptable than are false positives. To protect the public, stringent requirements are imposed on animal tests for potential carcinogenicity.

First, two species of animals must be used to test for carcinogenicity. Both sexes of each species must be used and there must be at least two exposure levels as well as a control group. In each cell of the experimental design, there must be a minimum of 50 animals. This leads to 600 animals required for each carcinogen test (2 [species] multiplied by 2 [sexes] multiplied by 3 groups [two treatment and one control] multiplied by 50 animals per group). To further improve the specificity of these assays, the strains of animals must have high rates of spontaneous cancers (highly susceptible to tumor development). As these spontaneous rates vary by cancer site, it is not uncommon to see animals tested for cancer at 20 or 30 body sites (at necropsy). This leads then to many, many comparisons between the experimental treatments and the control animals.

As shown in Table 10-3, the probability of finding a specific chemical to produce increased numbers of cancers in either species, in either dose group, or in either sex, is nearly random (46% chance of false positive). In practice, this is how the process is applied. If a chemical produces increased cancers in laboratory animals, it is considered a potential human carcinogen regardless of (1) whether there is a dose-response relationship, (2) which sex is effected, (3) which species is involved, or (4) even if the animals' tumors are in a site for which there is no human analog! Because at one time it was assumed that a threshold level did not exist for any carcinogen, a congressional statute "the Delaney Clause" now prohibits any chemical shown to cause cancer in laboratory animals to be used in human foods.

There is considerably more data on dose-response from animal experiments than for human dose-response data related to the ambient environ-

TABLE 10-3.   One Tail Probability of a False Positive Result*

Group	Sex	Cancer at any Site for Both Dose Levels	Cancer at any Site for Either Dose Level
Mice	Males	0.014	0.093
	Females	0.003	0.068
Rats	Males	0.013	0.178
	Females	0.023	0.235
Either Species	Either Sex	0.051	0.466

*Adapted from T. Fears, JNCI.

ment. Since the dose-response relationship is believed to provide the best estimate of risk, it follows that a person performing a risk assessment will choose to use animal data when human dose-response data are unavailable. Using animal data collected at high dosages (the animal studies), one can extrapolate backward (mathematically) to estimate lower dose levels in humans. However, for most risk assessments the extrapolation (extension of information beyond the range for which the data was collected) is highly questionable. Some of the questions associated with extrapolations are:

1. Does the animal metabolize large doses of the chemical exactly the same as it metabolizes small doses?
2. Do threshold (safe) levels actually exist?
3. Is human metabolism the same as animal metabolism, with different species, and with different chemicals?

Model Issues. Many mathematical models are proposed for the relationship of doses of a chemical to the production of health effects. The models are developed at the higher dose levels, but at the lowest exposure levels for which the models are extended they approach the intercept point at vastly different angles. Without human data to place in these models, risk estimates based on extrapolation from animals to man can vary greatly depending on the model employed in the estimation. These differences will translate into very different estimates of risk. For example, in Table 10-4 we can observe the variation in estimated risk of human bladder cancer from an ingestion of 0.12 mg/kg/day of saccharin, as estimated by the single hit model, the multihit model, and the Mantel-Bryan log-probit model. It is a separate question to ask, "Should the intercept point for an extrapolation model be

TABLE 10-4. The Estimated Risk of Developing Bladder Cancer From the Ingestion of 0.12 mg/kg of Saccharin per Day.*

Model Used From High to Low Dose	Expected Lifetime Cases per Million Exposed	Cases per 50 Million per Year
Rat dose adjusted to human dose by mg/kg/day equivalence		
Single-hit model	210	147
Multihit model	0.001	0.0007
Mantel-Bryan (log-probit model)	21	14.7

*Adapted from Upton 1988.

zero?" Much doubt exists whether exposure to a zero level of an agent is ever practically attainable. In the same sense, especially with cancer as the health effect of concern, there is an established background rate. Should (can) an extrapolation model be based on reaching a level of no cancer risk?

In any case, the objective of risk assessment is to estimate the probability of potential adverse health effects in relation to an environmental hazard. Extrapolation models are used for extending the data from high dosage levels (animal experiments) to lower dosage (exposure) levels (those of regulatory interest). Many models incorporate pharmacokinetics (metabolic) parameters for what is known about human physiology; usually these models are curvilinear (Figure 10.6). Other models incorporate parameters for the recognized detoxification and genetic repair capabilities humans have; these models are also nonlinear (Downs and Frankowski 1982).

Cancer is the primary health concern with risk assessment calculations. Much ado is made of the fact that cancer arises from a single cell (Yunis and Hoffman 1985). Because of this, the prevailing philosophy is that "each cellular injury works harm" (Meselson 1980). This single cell import leads to the "one hit" perspective, where it is generally thought that no safe exposure level exists (that is, no threshold)—there are only levels of minimized risk. Therefore, **risk assessments employ a linear (straight line), no threshold, extrapolation model.**

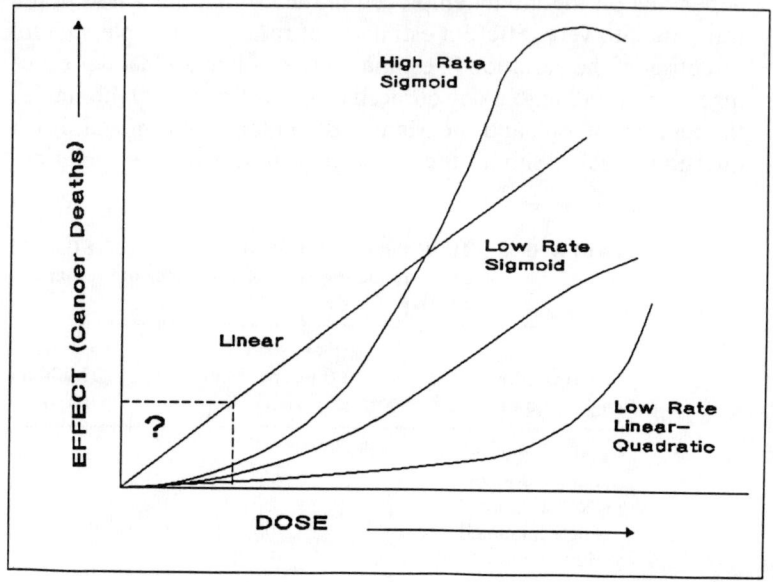

FIGURE 10-6.

Quantitative Methods. When we set out to do a risk assessment calculation, we must scale back down the dose gradient from the animal exposure levels. To do this, we use the models from ionizing radiation research (Glass et al 1990). The animal experiments provide the highest exposure data points on a regression model (for example, a graph of the dose of the chemical on to the rate tumors are produced, Figure 10.6).

It is a convention to consider all cancers as equals. This means that if the experimental animal gets cancers of a tissue for which humans do not have a counterpart (for example, the zymbal gland, a secretory gland in the rat ear) those cancers are nonetheless counted with the calculations of tumors produced. In screening chemicals for carcinogenicity, a comparison group of animals is studied to establish the expected rate of tumors for that species of animals in the experimental setting.

For a risk assessment calculation, place as many data points as are available, and considered acceptable, on the dose-response graph. A simple linear regression model is fitted to the data, with the constraint that the intercept be zero. With this model a slope relationship is established (beta); this slope then becomes the basis for the model extrapolation to the low dose levels for regulatory action.

For example, if 100 milligrams exposure of "tetraethyldeath" causes 10 percent tumors in rats, it is assumed that 10 milligrams of exposure will produce 1 percent tumors, etcetera. With this slope factor, the model is then applied to achieve an exposure level that will provide a risk of cancer of one-in-a-million. That is 0.000001. From the previous example with "tetra-ethyldeath," this would translate into 0.0001 milligrams of exposure. A sample regulation then might allow 0.0001 milligrams of "tetra-ethyldeath" as a safe public exposure. The subtleties of such calculations for air borne versus water borne agents is illustrated in the following calculations.

The convention of one-in-a-million cancer risk is often referred to as "ten to the minus six" due to the use of scientific notation (that is, 10^{-6}) when writing these values. More is said in the following page regarding the units in which risk is described. It is instructive to note that for occupational risk assessments, the "safe" level is considered 10^{-5}.

This difference is based on several considerations:

1. Workers can be more fully informed of the dangers from a particular agent, and directed to safe practices.
2. Workers can be monitored for compliance with safe practices.
3. Workers are compensated for their risks.
4. Workers are generally regarded as healthier than the general public (Monson, 1986).

Confidence Bounds. As with many mathematical and statistical endeavors, questions of precision arise. Precision is directly related to the sample size of the experimental groups, and is especially germane to the subject of extrapolating from animal to human species. Many of the available animal studies may have very few animals in them (occasionally only four or five animals per treatment group). To incorporate the potential error into the risk assessment calculations, 95 percent confidence limits are placed on each of the dose-response data points from the animal experiments. Then when the linear regression is performed to obtain the slope of the dose-response relationship, the upper 95 percent confidence limits for the animal data are used. Because of the small sample sizes, the confidence limits are quite wide. Some people criticize this practice because, with sparse data (that is, small sample sizes), the regulatory decision may be excessively conservative.

For example, with a small study, the wide confidence limits make for a higher, upper 95 percent confidence limit and thus a steeper slope for the dose-response model. Others argue that such sparse data situations are exactly where such precautions should be applied. Understandably the entire process of extrapolating from animal to man is quite controversial. The reader is directed to the recommended readings at the end of this chapter for sources of further information on this subject.

WORKING EXAMPLE OF THE RISK ASSESSMENT MODEL

We will use the chemical acrylonitrile for our example of a risk assessment procedure. Acrylonitrile is a water soluble, clear, colorless and highly flammable liquid with an unpleasant and irritating odor. Acrylonitrile is primarily used as raw material in the synthesis of acrylic fibers, styrene resins, nitrile rubber and barrier resins. Acrylonitrile (C_3H_3N) is also known as vinyl cyanide and is structurally "related" chemically to vinyl chloride. Carcinogenic concern led to a risk assessment for acrylonitrile.

A primary objective of risk assessment is to arrive at a determination of the "probability" of an adverse event occurring in an exposed population. "Probability" is expressed as the "unit" risk, which means the risk of having the adverse event occur per each exposure unit of the agent. One should note that ingestion exposures are presented in units of 1 μg/liter and inhalation exposures are shown as 1 μl/m^3. These exposure units correspond to the convention of parts per million (ppm). Also, one must become familiar with the use of scientific notation. Many of the formulae used with risk assessment are expressed in scientific notation. The jargon for risk assessment concepts reflects that use. For example, the "ten to the minus sixth" phrase is a representation of the one-in-a-million risk that is used for acceptable ambient environmental risk.

one-in-a-million is 1/1,000,000 or 0.000001 in scientific notation, 0.000001 is written as 1.0×10^{-6}

A sample risk assessment based on animal data: In the animal model, the formula for the slope of the hazard posed by acrylonitrile is:

$$B = p/d$$

where:

    B is beta, the slope of the regression line.
    p is the upper 95 percent confidence limit of the observed proportion probability of animals developing tumors exposed at the effective dose.
    d the exposure dose in milliliters of acrylonitrile / kilograms of the animal weight / day (ml/kg/day).

For a sample calculation, assume

$p = 5.0 \times 10^{-1}$

That is, 0.5 or 50% of the test animals (rats) will develop tumors at this dosage; this can be considered as the proportion of test animals that developed cancers from this exposure or the probability of cancer over the animal's lifetime occurring at this exposure.

d = 1 ml acrylonitrile per liter of water / day for test animals
    = 1 ml/l/day that the animals drank at liberty.

To translate this animal exposure into human terms, 2 ml would be contained in the 2 liters of water that a normal 70 kilogram person drinks each day. A regulatory action then aims for units of exposure in parts-per-million, or ppm; in water, this is μl/l. Two milliliters would be 2000 μl divided by 70 kg, this means that d for the "normal" person would be 28.6 μl/kg/day.

Therefore B (the slope of the linear dose-response model) becomes:

$B = p/d$ = 0.5 / 28.6 μl/kg/day
B = 0.01748 (lifetime probability of cancer) / μl/kg/day for a lifetime.

B is the slope for the low-dose extrapolation from a "lifetime" daily dose (d) of 28.6 μl/kg/day. Remember this extrapolation is not to be used for doses greater than 28.6 µl/kg/day. Since this is a safe dose extrapolation to an intercept value of zero, we can apply our derived results to a "lifetime" risk of cancer from this daily exposure.

$$\text{the probability of cancer} = (0.01748)\,(d)$$

This slope extrapolation translates to "17,480 cases of cancer per million persons if they were exposed to 1 ppm of acrylonitrile in their drinking water for their entire (70 year) lifetime." This probability of cancer is 17,480 times to high for regulatory criteria of one-in-a-million risk of cancer. This value may be revised by setting the acceptable risk to one cancer case per million and dividing by 17,480.

$$1 \text{ cancer} / 17{,}480 / \mu l/kg/day = 0.000057 \ \mu l/kg/day/\text{lifetime}.$$

This means that a regulatory action would require that acrylonitrile levels in public drinking water must not exceed 0.000057 ppm (or 57 ppt [trillion]) to be regarded as safe.

In some risk assessments human data are available to incorporate into the risk analysis. Such data are almost exclusively occupationally derived: This is the only setting where exposure monitoring makes available the necessary dose-response information. In the case of acrylonitrile, when the original risk assessment was completed in 1982, there were ten occupational epidemiologic studies available. However, nine of the studies were excluded for various reasons and only one study was used (O'Berg 1980).

O'Berg's study (1980) was an occupational cohort study of the employees at a DuPont Chemical Plant manufacturing textile fibers. The "exposed" cohort was composed of 1,345 men who had been employed at the plant any time during the interval from 1950 through 1966. Follow-up for these men was focused on the decade 1967–1976. Using various data sources (including DuPont's own cancer registry), ascertainment of cases was 98 percent. No information was available on exposures during 1950–1966, but a panel of industrial hygienists scored the workers for having high, medium, and low exposures based on job title and functions. These exposure levels were estimated from monitoring levels for various jobs since 1970 as 20, 10, and 5 ppm respectively. The average duration of employment in these "exposed" jobs was 9 years. The observed cancer rates were compared to employees from other DuPont plants and were divided into salaried and wage pay categories. The greatest increased cancer risk was for "wage" category employees. There were 25 cancer deaths in the entire cohort, with 20.5 cancer deaths expected. There were eight lung cancer deaths, with 4.4 expected. No smoking history was available on the workers.

Here is a sample risk assessment based on human data. In the human model, the formula for the slope of the hazard posed by acrylonitrile is:

$$\beta = \frac{P_o(R-1)}{x}$$

where:

- β is beta the slope of the regression line.
- P_o is the probability of dying with cancer before age 60 years (because of the average age in O'Berg's cohort (1980)).
- R is the level of risk associated with the exposure level selected.
- x is a weighing factor for translating an occupational exposure into a representation of constant, lifetime exposure. For example, this occupational exposure is adjusted to a continous "lifetime" exposure based on 8 hours out of 24 hours for the working day; 240 out of 365 for the days worked per year, and 9 years out of 60 years for the average length of employee exposure.

From the O'Berg study (1980), the lung cancer risk for wage employees with the high and median acrylonitrile exposure and the longest follow-up period were 5 lung cancers with 1.4 expected: a risk of 3.6 times the expected. The exposure level for this group was assigned to be 10 milligrams of acrylonitrile/cubic meter of air (10 ppm) per day (mg/m^3/day), e.g., the middle category of exposure in the occupational group.

For a sample calculation, assume:

- P_o = 0.036 (the cumulative lifetime risk of cancer to age 60)
- R = 4.0 (rounding off the 3.6 Standardized Mortality Ratio from O'Berg's study's results - this is for lung cancer).
- x = 10 ppm (8/24) (240/365) (9/60)
      = 0.329 ppm/day/yr for a "lifetime"

now calculate the slope (B) for the linear dose/response model:

$$B = \frac{(0.036)(4-1)}{0.329 \text{ ppm for a "lifetime"}}$$

(cumulative lifetime risk multiplied by the excess risk due to exposure).

B = 0.328 lifetime probability of cancer / ppm / "lifetime"

This slope translates into 328,000 cases of cancer per million persons if they were exposed to 1 ppm of acrylonitrile in their air for their entire (60 year) "lifetime." This value is 328,000 times too high; it can be revised to the acceptable risk of one-per-million by dividing the one case-per-million by 328,000.

$$1 \text{ case} / 328,000 = 0.000003 \text{ ppm/lifetime.}$$

This means that a regulatory action would require that acrylonitrile levels in public air must not exceed 3 parts-per-trillion 10^{-6} mg/m^3 to be regarded as safe.

Discussion of Findings: It is noteworthy that the results from the animal data led to a regulatory level of approximately 57 ppt of acrylonitrile in water, and the human exposure data provided an acrylonitrile level in air of 3 ppt. *[Please notice that these data are taken from some earlier work with acrylonitrile and should not be viewed in terms of contemporary exposure limits, they are used here simply as a model for the purpose of demonstrating these extrapolation calculations.]*

Regulatory action at the ppt level suggests that acrylonitrile is a potent carcinogen. Note that the test animals developed cancer of the zymbal gland, and the humans developed cancer of the lung, with no consideration given to smoking history of the subjects. The 3.6 RR for lung cancer among these employees is about one third that for a heavy smoker.

HOW TO USE A RISK ASSESSMENT

With the extrapolated dosage that is obtained in performing a risk assessment calculation, standards are set for ambient exposures to a particular toxic agent, whether it be an occupational or ambient exposure. In this process there are many value judgments that must be weighed, including those cited regarding the specific health effects associated with exposure (carcinogens receive much greater regulation than, say, an agent producing dermatoses). The difference between ambient and occupational exposures was previously discussed. This "where" aspect of exposure leads logically to the issue of "who" is exposed. Should a regulatory decision be made on the basis of heavily exposed persons and/or highly susceptible persons (children, the elderly, the sick)?

An important aspect of developing a risk assessment is understanding the way that risk associated with exposure is described. The most critical description is called **unit risk**. Unit risk is based directly on the slope of the regression model, and refers to the increased health risk per unit of exposure (that is, with the "tetraethyldeath" example—one milligram of exposure would be equal to 0.1 percent increase in the lifetime rate of cancer).

Critical to understanding the proposition of unit risk is recognizing that it is based on the assumption of lifetime exposure. Generally, the "unit" risk is based on 70 years exposure, and not on an instantaneous or one time exposure. Another common description for risk is the **maximum individual risk (MIR).** The MIR is directed toward the persons most likely to receive the heaviest exposures.

Some examples of how this concept might be applied relate to children and the pesticide Alar (a ripening agent used on apples). Children might represent the most heavily exposed persons, so the **maximum daily intake (MDI)** would be calculated for their exposure. This exposure would then be compared to the exposure derived for the "general public," the unit risk based on a lifetime of 70 years of exposure. For computing these different measures, assumptions are made about the MDI. With Alar, the assumption was that adults do not drink as much apple juice as children, thus their exposure would be less. The MDI estimate on children would be compared with the unit risk estimate resulting from the 70 year "lifetime" exposure. A risk management decision will be made, either on the basis of the unit risk, or on the MDI.

In risk management (more information is provided on risk management issues in Chapter 11), **worst case** reasoning is used. Let's consider the risk manager's response to Alar and apples. In the worst case scenario risk management is not based on factors such as "average" consumption of apples, or for usual usage. Apple consumption in children is used to establish an MDI for calculation. In many risk assessments, pregnant women are used to establish the "worst case" scenario because the fetus is highly susceptible to environmental insult. Sometimes older persons, cigarette smokers, or AIDS patients, etcetera may be used to establish the "worst case" scenario. In other words, to establish an MIR and an MDI, the most vulnerable and susceptible people in the exposure process are used as the standard. Often, the basis upon which these people are picked is very subjective (that is, a child three months old, or one a year old, etcetera). In the worst case scenario, as with the selection of animal experiments to include with the risk assessment, one hopes to develop a **best estimate** of exposure.

Interpreting Results

If this discussion of **worst case** and **best estimate** leaves you feeling a bit uncertain about using risk assessment for a regulatory decision, we have accomplished our goal. A healthy **skepticism** must be applied to risk assessment, as well as to risk management. You must, however, also keep in mind that the responsibility of a regulatory agency is to **regulate,** and this requires the best information possible on potential risks to the public.

A few final precautions for interpreting a risk assessment calculation follow below:

1. Risk is **probabilistic.** Thus, the results are based on the probability of an adverse health effect taking place in a particular population. The effect, in that population, is not an absolute certainty to take place.
2. Risk is presented as an **incremental** increase among exposed persons. Meaning that there is an expected rate of incidence of the event. In the case of cancer, with a lifetime background risk of 30 percent, that means that there are to be expected 300,000 cancers per million people. A risk management decision based on one case per million people will increase the risk of exposed persons to 300,001 cases per million people over a lifetime.
3. Risk assessment is a **predictive** concept. Risk assessment is based on the estimation of a specific exposure, to a defined population, yet the findings are extrapolated to the total population including persons that are highly susceptible to environmental insult (for example, smokers, persons with immune deficiencies, family histories of disease, older persons,) and may not be, at all, representative of the highly selective exposed population.
4. Risk assessment is **not presented for the individual.** Risk assessment is a population concept, and that makes epidemiology relevant to the process.

SUMMARY

There are legitimate concerns about the use of epidemiologic data in risk assessment. Clearly, there are potential difficulties relating to:

1. Multichemical exposures,
2. Specificity of exposure-effect,
3. Sensitivity of exposure-effect,
4. Applicability of exposure assessment methods to epidemiologic study designs,
5. Data quality,
6. The appropriateness of the study design,
7. The power of the study to detect a significant difference in the populations being studied,
8. Intervariability (between groups and persons),
9. Intravariability (within groups and persons),
10. The issue of biologic susceptibility for a particular agent as an initiator or promoter of disease, and
11. Finding effects with low dose exposures that are usually found in epidemiologic studies.

Risk assessors must also be concerned about potential confounding and selection bias, and extrapolating from animal studies and the associated high dose levels, across species, to low-level risks in human populations. Risk assessor's must also understand and work with temporal relationships, disease latency, and the duration or periodicity of exposure.

In this chapter we have attempted to provide an overview of the elements, and processes involved in risk assessment (for example, using human data, animal data, mathematical extrapolation models). We have discussed the decision process in performing a risk assessment with a **real world model** as an example. Lastly, an introduction to the interpretation of risk assessment findings has been offered. In Chapter 11, we will discuss the risk management side of the process, which includes interpreting risk assessment findings, and communicating risk concepts.

ASSIGNMENTS

1. Calculate the unit risk (1 mg/l) for oral exposure (assume an "average" 70 kg person consumes two liters of water a day). For p use 1.7×10^{-1}/mg/kg/day.
2. Calculate the unit risk (1 mg/m^3) for inhalation exposure (assume an "average" 70 kg person respires 20 m^3 of air a day). For p use 4.2×10^{-1}/mg/kg/day.
3. Calculate the unit risk (1 mg/l) for inhalation exposure using the epidemiologic data and a suspected cancer risk of 8.0 and an exposure level of 5 ppm.
4. Decisions related to the use of risk assessments are highly controversial. Consider formaldehyde. The safe exposure level was found to be higher than the exposure standard at that time (that is, the regulatory standard was less than the suggested safe level). Why would special interest groups oppose a decision not to regulate formaldehyde in the face of such information?
5. Faced with this perspective on carcinogen testing, consider the concept of pseudo-carcinogencitity (Roe 1983). This proposition submits that control animals gain weight from overeating and are at higher risk of some cancers because of obesity. This would impact on the potential direction of random errors in carcinogen testing. What is the direction of this effect?
6. Benzene was regulated at 10 ppm. A proposed reduction to 1 ppm was opposed by industries as being unable to be achieved without prohibitive costs. Why was the compromise decision of 5 ppm inappropriate?

RECOMMENDED READING

Gordis, L. 1988. *Epidemiology and Health Risk Assessment.* New York: Oxford University Press.

Paustenbach, D. J. 1989. *The Risk Assessment of Environmental and Human Health Hazards: A Textbook of Case Studies.* New York: Wiley and Sons.

REFERENCES
Allman, W. F. 1985. "We have Nothing to Fear (But a Few Zillion Things)." *Science* 85, 38–41.
Ballantyne B. 1985. "Evaluation of Hazards From Mixtures of Chemicals in the Occupational Environment." *J Occupational Medicine* 27 (2):85–94.
Bender, A. P., Williams, A. N., Johnson, R. A., and Jagger, H. G. 1990. "Appropriate Public Health Responses to Clusters: The Art of Being Responsibly Responsible." *American Journal of Epidemiology* (132):S48–S52.
Downs, T. D., and Frankowski, R. F. 1982. "Influence of Repair Processes on Dose-Response Models." *Drug Metab. Rev.* 13:839–52. Downs, T. D. 1985. "Assessment of Various Dose-Response Models in the Determination of Risk." In: *New Approaches in Toxicity Testing and Their Application in Human Risk Assessment.* ed. A. P. Lee. New York: Raven Press.
Enterline, P. E., DeCoufle, P., and Henderson, V. 1973. "Respiratory Cancer in Relation to Occupational Exposures Among Retired Asbestos Workers." *British Journal of Industrial Medicine* 30:162–66.
Glass, L. R., Jones, T. D., Easterly, C. E. and Walsh, P. J. 1990. *Use of Short-term Test Systems for the Prediction of the Hazard Represented by Potential Chemical Carcinogens,* ORNL TM-11413. Pub. Oak Ridge National Laboratory, Oak Ridge, TN.
Glasser, J. H. 1985. Health Statistics Surveillance Systems for Hazardous Substance Disposal. Proceedings from the *1985 Public Health Conference on Records and Statistics,* DHHS Pub. No. (PHS) 86-1214, pp. 221–24.
Goldsmith, J. R., and Needleman, H. L. "Lead Exposures of Urban Children: A Handicap in School and Life?" In: *Environmental Epidemiology: Epidemiological Investigation of Community Environmental Health Problems.* Boca Raton, Florida: CRC Press, Inc.
Griffith, Jack., and Duncan, Robert C. *National Monitoring Study: Citrus. (Vol. II). An Assessment of Grower Pesticide Use Practices in the Florida Citrus Industry.* University of Miami Press. 1983.
Hammond, P. B., and Beliles, R. P. 1980. "Metals." In: *Casarett and Doull's Toxicology.* eds John Doull, Curtis, D.
Klaassen, C. D., and Doull, J. 1980. "Evaluation of Safety: Toxicologic Evaluation." In *Casarett and Doull's Toxicology,* eds J. Doull, C. D. Klaassen, and M. O. Amdur. New York: Macmillan Publishing Co., Inc., p. 21.
Klaassen, and Mary O. Amdur. 1980. "Evaluation of Safety: Toxicologic Evaluation." In *Casarett and Doull's Toxicology,* eds J. Doull, C. D. Klaassen, and M. O. Amdur. New York: Macmillan Publishing Co., Inc.
Monson, R. R. 1986. "Observations on the Health Worker Effect." *Journal of Occupational Medicine* 28(6):425–33.
Meselson, M. 1980. "The Story of Agent Orange." *National Geographic* 157(2)145:82.
National Research Council. 1983. "Commission on Life Sciences. Committee on the

Institutional Means for Assessment of Risks to Public Health." *Risk Assessment in the Federal Government: Managing the Process.* Washington, DC: National Academy Press.

O'Berg, M. T. 1980. "Epidemiologic Study of Workers Exposes to Acrylonitrile." *Journal of Occupational Medicine* 22(4):245–52.

Shoedell M. 1985. "Risky Business." *Science* 85 43–47.

Stallones, R. A. 1988. "Epidemiology and Environmental Hazards." In: *Epidemiology and Health Risk Assessment.* (ed.) Gordis, L. Oxford University Press, New York, Oxford, pps 3–10.

Roe, F. J. C. 1983. "Testing for Carcinogenicity and the Problem of Pseudocarcinogenicity." *Nature* 303:657–8.

Upton, A. C. 1988. "Epidemiology and Risk Assessment." In: *Epidemiology and Health Risk Assessment.* (ed.) Gordis, L. Oxford University Press, New York, Oxford, pps. 18–36.

U.S. Department of Health, Education and Welfare. 1964. "Criteria for Judgement." In: *Smoking and Health. Report of the Advisory Committee to the Surgeon General of the Public Health Service.* Washington, DC: U.S. DHEW, Chapter 3.

U.S. EPA. 1979a. *Chlorobenzilate Position Documents 3. and 4.* Office of Pesticide Programs, Office of Pesticides and Toxic Substances.

U.S. EPA. 1979b. *Information Taken from Suspended and Canceled Pesticides.* Office of Public Awareness (A-107), Washington, D.C. 20460.

Yunis, J. J., and Hoffman, W. R. 1985. "Birth of an Errant Cell: A New Theory About the Cause of Cancer." *Minnesota Magazine—(Special Issue) The Sciences* Nov/Dec:28–33.

11
Public Communication, Participation, Risk Management

Tim Aldrich, Jack Griffith, Robert Gustafson, David Graber

OBJECTIVES

This chapter will:

1. Discuss prevalent public attitudes in response to issues involving choices for environmental actions.
2. Identify focal issues and perspectives that impact on the public's lack of regard for government agencies related to environmental actions.
3. Describe advantages and disadvantages of working with community organizations for conducting an environmental epidemiologic study.
4. Explain the concept of Earth as a spaceship.
5. Discuss three fundamental issues in risk management and public participation: The basis upon which a regulatory agency decides to assess the risk associated with exposure to a specific substance; the kinds of evidence, scientific and economic, required by a regulatory agency in the risk management process (see Chapter 10 for specific data types); and the risk management strategies to be used by the regulatory agency.

PUBLIC COMMUNICATION

"Public opinion is everything. With public sentiment, nothing can fail. Without it, nothing can succeed. Consequently, he who molds public opinion goes deeper than he who enacts statutes or pronounces decisions." Abraham Lincoln

Effective risk communication may be accomplished via public hearings or small group discussions of key individuals. The media can participate in the dissemination of important information through prepared statements with carefully planned content. Whenever possible, we should try to work with informed media representatives (for example, scientific writers). Cultivating credibility before there is a "crisis story" is also a sound policy. We should strive to keep lines of communication to the media open at all times, not just when there is an emergency situation.

Elements of Effective Communication

The EPA (1987) has developed several fine booklets on the subject of risk communication. From these sources, seven elements of effective risk communication have been selected:

1. Accept and involve the public and health officials as legitimate partners to the risk communication effort.
2. Plan carefully what you want to communicate (before) and evaluate how well you communicated (after).
3. Listen carefully to both the words and the underlying concerns of persons addressing you.
4. Be honest, open, and frank.
5. Coordinate and collaborate with the public and health officials in the process of risk communication.
6. Understand the role of the media and communicate with them.
7. Communicate clearly (avoid jargon) and with compassion.

The media represent an important part of the risk management, and risk communication, process. However, the media operates from an entirely different perspective than scientists. Very often, you will find that:

1. Reporters will have limited time (for example, pressing deadlines) and expertise. Many larger newspapers, television stations, etcetera will have science reporters with a deeper grasp of basic scientific principles (for example, statistical uncertainty), but in practice this is generally not the case.
2. Reporters are frequently just as interested in opinion as they are facts.
3. Claims of risk are more newsworthy than claims of safety.
4. Pour environmental risk into a pot, and stir with a liberal seasoning of politics, and you have a potentially hot media item.
5. Many consumers of mass media prefer simple dichotomies (for example, yes there is a health risk, no there is not) to accurate, yet technical, expla-

nations that appear equivocal. A reporter, in search of a story, may pressure the scientist to oversimplify an issue, or answer with a simple yes or no.
6. Attempts are made by the media to personalize risks. This is often accomplished with human interest stories. Invoking the aspect of tragedy can be a major part of developing the human interest angle. Also, the man on the street view, while very popular, is not a substitute for informed reporting.
7. When the risk management process is completed, provide complete disclosure of the findings from the risk assessment. Be forthright in identifying uncertainties. Give credit to the legitimate concerns of the public and be compassionate for their fears. Where possible, use a credible local representative to discuss the more subtle issues and to ease the tension associated with the "us versus them" dynamics.

COMMUNITY ACTION AND INTERACTION WITH HEALTH AGENCIES

Local observers often notice cancer clusters or environmental hazards before scientists and officials. A suspected leukemia cluster was brought to the attention of officials by residents of Woburn, Massachusetts (Brown 1987). A group of concerned women brought the 2,4,5-T exposure and miscarriage issue described in the following pages (251–2) to the attention of the EPA. Cancer clusters in the work place are typically reported by employees (Frumkin and Kantrowitz, 1987). Through conversations and personal observation, citizens may come to suspect unusually high cancer occurrence in their locality. At or about this time, concerned individuals may form local organizations to address the perceived problem.

The quality and sensitivity of the health agency response to the initial report of a hazard often sets the tone for future interactions. Among individuals living adjacent to a landfill, "...basic distrust was heightened because of a lack of responsiveness by the investigating agencies to requests for information from the homeowners" (Leitko 1985). In our experience, poor and inconsistent responsiveness to citizen requests by government agencies is always a primary source of anger and mistrust.

With a disease cluster report, an initial response should always include an offer to travel to the site and meet personally with local residents. Personal interaction with concerned residents helps develop rapport, and allows the health agency to be receptive to their concerns. At public meetings, it is possible to provide education on such topics as risk factors, latency, and environmentally related diseases.

Issues in Working With Community Groups

The past decade has witnessed a number of incidents where government health agencies and scientists were viewed as:

1. Unconcerned about community health (Krauss 1989)
2. Tending to minimize or discount the potential threats to community health (Reich 1983)
3. Unwilling to shoulder the responsibility to correct a hazardous situation (Harris 1984)
4. "In league" with big business and industry (Masterson-Allen 1990; Brown 1987)

A survey of community action groups showed that 45 percent believed that state or federal agencies had obstructed their access to data (Freudenberg 1984). The health official investigating a cancer cluster or environmental hazard, and who expects to be regarded as a benevolent friend of the people, may be in for an unpleasant surprise. An honest commitment to scientific integrity is often misinterpreted. Harris (1984) has aptly stated:

1. Underlying much of the problem is the public's intolerance of ambiguity, their yearning for simple declarations.
2. Health officials punctilious concern about the inadequacy of scientific evidence and their disinclination to draw conclusions from insufficient data are easily mistaken for lack of resolve or abdication of the responsibility to act.

We have observed considerable variety among concerned citizens groups in the United States. At first, most are not in an adversarial position with health officials, but it is not unusual for this to change after a series of negative experiences in which their concerns are repudiated or are not taken seriously. Awareness of potential sources of antagonism at the outset of an environmental epidemiologic investigation is invariably valuable (Lynn 1986).

Statistical Versus Practical Significance

In meetings with community groups, an explanation should be offered that completely expains the objectives of the environmental investigation. Such concepts as "the spatial proximity of cancer cases," should be fully explained. Importantly, the meaning and importance of "statistically significant" must be communicated. The widespread acceptance and use of the .05 significance level places a fairly rigorous premium on avoiding a Type-I error (viz., accepting a false [alternative] hypothesis). This standard may be acceptable for general scientific and academic research, but it is extremely conservative when the health of a community is at stake. In epidemiologic studies involving small areas or populations, use of the .05 level will generally preclude finding statistically significant health effects. Therefore, it is critical to distinguish between public health significance and statistical significance (Ozonoff and Boden 1987).

Cad (1986) observes:

> "The evidence that epidemiologists require to achieve scientific statements of probability exceeds the evidence required to state that something should be done to eliminate or minimize a threat to health. The degree of risk to human health does not need to be at statistically significant levels to require political action. The important political test is not the findings...of the nonrandomness of an incidence of illness but the likelihood that a reasonable person...would avoid the risk."
>
> "In investigating disease clusters that may be linked to environmental hazards, the public health scientist may have to depend on personal judgement to decide if the traditional significance measures are appropriate in evaluating a health threat to a community. She may be placed in an ethical dilemma. A liberal acknowledgement of a potential disease cluster may threaten her scientific credibility. On the other hand, a rigid adherence to a scientific "formula" for significance may alienate the community. In any event, community residents should be informed that when a result is said to be 'not statistically significant,' this does not imply the automatic and concomitant dismissal of what they perceive to be a serious health problem."
>
> "In situations where the incidence appears high, but fails to achieve statistical significance, it is prudent to maintain public health surveillance on the area's morbidity and mortality, on a yearly basis. While this is not always the optimum solution, the error of permanent cessation of oversight is avoided."

THE JOINT AGENCY AND COMMUNITY EPIDEMIOLOGIC STUDY

Ozonoff and Boden (1987) contend that the most appropriate response to pressures from local groups is to:

> "...involve local citizens in the study from the design stage onward..." Their involvement will also educate them about the difficulties and uncertainties of performing and interpreting these studies. All studies should also include a plan for informing the wider community of the progress of the study..."

Local citizens have proven to be valuable partners in cancer cluster investigations. When possessed of energy and enthusiasm, they can expedite study progress. In rural areas, their knowledge of local conditions and people invariably results in time saved and many unnecessary problems avoided. By living "on-site," they can be much more efficient than health officials from a distant city. Local participation serves to raise the status of the group, increases their contribution of "valid" material, and also increases the likelihood they will support the findings (Leitko 1985). An effective community group, formed to address an asbestos exposure hazard, resulted from the efforts of a labor union, a corporation, community health care providers, and a medical school (Holstein et al. 1984).

Joint disease cluster investigations, involving lay persons, have been criticized for being unscientific. Often local individuals have a stake in the outcome (that is, they hope and expect to uncover and identify a health hazard). This lack of objectivity may contribute to bias in overreporting cancer cases. During preparatory meetings, prior to the survey, epidemiology staff may train local workers in such areas as accuracy of completed information, confidentiality, and problems of interviewer bias.

The joint completion of the study may result in a greater respect for, and awareness of, health agency and local positions by both groups. Public health agencies may successfully work together with local residents and citizen's groups even though possessed of different beliefs and perspectives at the outset. To accomplish this, a new egalitarianism between scientist and layman must evolve. The scientist, who will be working with local residents in projects relating to disease clusters or environmental hazards, would be well-advised to display empathy, develop human relations skills, and be aware of possible conflicting values and goals between herself and the community group. Joint projects, if carefully planned, can bring health agencies and concerned citizen groups closer together and be scientifically fruitful (Lynn 1987).

THE NORTH CAROLINA RULE

In rural Scotland County, North Carolina, construction of a processing plant for treating hazardous waste was opposed by local citizens. The focus of the contention was the discharge of processed waste water into the Lumber River. The publicly owned treatment works for a town downstream processed 500,000 gallons of water per day. The contracting company proposed to discharge 500,000 gallons of waste water based on this capacity. North Carolina statutes require a 1,000:1 dilution factor for industrial discharges into public waters. This limitation would have curtailed the discharge level to 72,000 gallons per day, a level that the owners of the proposed hazardous waste treatment processing plant did not find economically feasible. In

November 1987, EPA Region IV initiated proceedings to rescind the Resource Conservation and Recovery Act (RCRA) authority for the state of North Carolina to manage its own hazardous waste.

In April 1990, an administrative law judge for the U.S. Environmental Protection Agency ruled that this restriction by the state of North Carolina did not infringe on the economic options for the company to construct a similar plant elsewhere, or a smaller plant in Robeson County. The company filed a countersuit against the state of North Carolina alleging that the state had acted inconsistently with federal regulations under RCRA. Three months later an EPA regional director upheld the ruling in favor of the North Carolina decision, citing Section 3009 that states: "Nothing in this title shall be construed to prohibit any state or political subdivision thereof from imposing any requirements, including those for site selection, which are more stringent than those imposed by such regulations (Conservation Council 1988).

This decision, sometimes termed the "North Carolina Rule," was a major victory for local citizen action groups opposed to perceived environmental threats. This experience points up the process involved with many environmental actions. Local chamber of commerce representatives had initially sponsored the hazardous waste treatment plant's construction on the basis of economic growth for the local economy (jobs, revenues from taxes, etcetera). Opposition to the citing developed quickly. No less than seven community groups banded together to oppose the action by alerting, informing, educating citizens not only about the proposed hazardous waste facility but about the importance of informed and effectively organized citizen involvement in public policy issues.

This community action cannot be characterized as an example of the concept of NIMBY (Not In My Back Yard) if by that concept is meant unthinking and ill-informed opposition to the citing of a hazardous waste facility. Opponents to the citing had sought out and used experts whose studies supported not citing the plant in Scotland County.

It is fair to say the community action was characterized by a reverence and ethos for the earth as a self-contained and limited resource, a "spaceship." While the spaceship metaphor was not used by this community action group, it does reflect their feelings. The spaceship metaphor arises from a quote from the economist Kenneth Boulding:

> "We have been able to regard the atmosphere and the oceans and even the soil as an inexhaustible reservoir from which we can draw at will and which we can pollute at will. There is writing on the wall, however...even now we may be doing irreversible damage to this precious little spaceship."

Far from encouraging a moratorium on technological development, this perspective seeks to affirm that technological progress must move speedily ahead in order to control exhaustion of resources and increased pollution. However, this progress must operate"...within the context of an ethic of parsimony rather than exploitation." (Shinn 1972). It is imperative that individuals within communities must make choices regarding environmental actions and that the political process must be invoked for the eventual decisions. Political decisions may take the form of social accountability that reckons the cost of environmental damage and ascribes that to the industrial source for associated pollution (Ruckelshaus 1983). A factory must reconcile the costs for pollution impact versus **mitigation** of that pollution; social accountability compels this action (Shinn 1972).

The Environmental Ethic

However, it must be recognized that there are forces operating in communities to promote economic growth who are also in favor of environmental ethics. These perspectives come together in what has been termed the "BOTH" movement. Faced with closing a copper smelter for emission violations, a community sought to keep this economic "center" for their community while preserving a more healthful environment (Ruckleshaus, 1984). These issues of economic gain and community harm have raised suggestions of environmental inequity for minority and economically disadvantaged populations (Bullard 1990). These implications are substantive concerns and are clearly fundamental aspects of an environmental ethic.

An environmental ethic must, by definition, focus on the environment: What protects and enhances the environment as well as what threatens or destroys it. A sense of personhood cannot be ignored, for the value of the individual and the rights associated with personal freedom are important considerations. An attitude that people are stewards of nature and the natural order is espoused. Nature should be advocated as a dynamic entity, evolving and creating through its evolution. Nature must be recognized as having its own potentials and limits. The interdependence of living things is a central theme to the environmental ethic. The holistic view of life presents a unified ecosystem with diversity and vulnerability, balanced within the narrow limits of gravity, heat, atmosphere.

Earth is a unique spaceship with air, water, and soil as resources for its passage. It is a cliché to say that we live in a technological age and that we must learn to control it if we are to survive. Equally unacceptable is the claim that as a trade-off for living in an age of technology, citizens are obligated to live near hazardous conditions. The approach followed in North Carolina involving informed scientific, technological, and citizen participation, responded to the needs of handling hazardous wastes while protecting its citizens.

THE PUBLIC HEALTH FUNCTION

Public Health Agencies

Public health agencies continually receive requests to investigate cases of disease in a community. Very often, these situations involve only a few scattered cases, or in some instances, a small clustering of cases. Occasionally, but not always, the agency will send an investigator to evaluate the merits of the potential health problem. Clearly, the people in the community believe it's a problem or they wouldn't have asked the agency to intervene. The investigator's task is to evaluate the cases, and make a determination as to the etiology of the disease. To do this, the investigator must first determine whether the cases are just an unfortunate random event, or if the cases are somehow linked to exposure to an environmental toxicant.

Frequently, the investigator can find no evidence that the cases and the exposure are related. Occasionally, the investigator will suspect a linkage between exposure and effect, and a full-blown environmental epidemiologic study will be undertaken. Developing an epidemiologic study during a time of intense public scrutiny requires that the investigator work closely with citizens of the community. In fact, community involvement in many instances is so acute that studies such as these tend to represent a resurgence of old fashion shoe-leather epidemiology that was common during the early years of infectious disease epidemiology (Warner and Aldrich 1988; Fiore et al. 1990).

Today however, the interest of the public tends to focus on selected chronic (for example, cancer or neurological diseases), or reproductive diseases (for example, birth defects, miscarriages). The study of the etiology of chronic diseases is much more difficult for the epidemiologist, since there is no easily predictable incubation period or easily diagnosed symptomatology. In the study of chronic diseases, the epidemiologist must deal with health effects that have subtle and complex biological manifestations, and very complex multimedia exposures. To complicate matters further, exposures in the community are usually at or below detection limits, which creates another formidable problem when attempting to assess personal and community risk.

In the Chapter 1 discussion of some recent environmental events, we described situations in which the public believed it was suffering adverse health effects related to environmental exposures. Despite the best face that the health agencies could put forward, the public demanded and received a resolution to the exposure problem. Interestingly, at no time in any of the events depicted in Chapter 1 was there ever a causal relationship developed between a disease and an exposure.

People believe, intuitively perhaps, that exposure to chemicals is simply not healthy, and subject to economic constraints, they prefer to have it reduced or eliminated. In each of the events discussed in Chapter 1, the public did not accept the claim of the authorities that any risk associated with the exposures was minimal. In fact, the public's perception was that the authorities did not understand the magnitude of the problem, or that they were "covering up." The difference between the public's perception of the problem and that of the health authority has been likened to a credibility gap. Now that public anxiety over environmental contamination is reaching new heights, epidemiologists are faced with methodologic and ethical challenges as they grapple with these "environmentally associated diseases." (Gough 1987; Goldsmith 1988).

The credibility gap appears to be growing as more and more information on health risks, primarily generated from animal studies, is brought to the attention of the public by the various news media. The health professional must also take a share of the blame in promoting the gap. Recall, in Chapter 1, the Love Canal residents were told by the EPA that people tested in the community were experiencing excessive rates of genetic mutations. Later, upon further review, an EPA advisory group denounced the findings, citing a severely flawed study design. This review was too little and too late for the residents of Love Canal. They demanded that the government purchase their property, and that they be permitted to move elsewhere. Interestingly, even if the results of the study had been unimpeachable, no causal association would have been possible. Wolff (1983) has suggested that damage to genetic material has not yet been shown to be predictive of ill health. Although a population with damaged chromosomes may have more cancer and birth defects than expected, chromosome damage is simply an indicator that the population was exposed to a DNA toxicant, and is not a "harbinger of cancer or birth defects."

The use of biomarkers (for example, sister chromatid exchanges, or chromosomal aberrations) in epidemiologic studies, for either exposure or effect, must be explained to the public in a forthright and understandable fashion. Clearly, the Love Canal residents with the supernumerary acentric chromosomal aberrations could not be told, scientifically, that their risk to cancer had increased because of the mutations. Yet, that was the implication. Clearly, the credibility of scientific results is a substantive issue in the acceptance by the public of risk assessment findings and risk management decisions.

The Political Gap in Risk Management

The public has grave misconceptions regarding scientific measurement and acceptable risk (Ozonoff and Boden 1987; Cad 1986). Epidemiology assumes a baseline level of [absolute] risk, whereas the public is interested in zero risk (Cox and Ricci 1989, Gordis 1988). Scientific equanimity with regard to the expected occurrence of disease draws public criticism concerning lack of objectivity (Cad 1986). There is also political, as well as public, resistance to accepting that disease will occur (Brownlea 1981). As Ruckleshaus (1983) suggested, these fundamentally different positions cry out for the scientific community and the public to transcend personal bias and accept the premise that health is a right and illness a statistical certainty.

As a case in point relative to this evolving reconciliation, lets consider asbestos. Asbestos has been at the forefront of many environmental cancer debates and has been central to compensation litigation (Chase et al. 1985). The following is a prototype environmental hazard that will demonstrate the extent of this schism in attempting to develop an appropriate risk management decision:

1. Asbestos exposure is widespread.
2. The referent human health risk (mesothelioma) is quite rare.
3. Laboratory and human risk assessments have been developed and debated.
4. Confounding by lifestyle factors [smoking] is great.
5. The risk level is consistently found to be small, and recognition of the hazard is greatly removed from the beginning of exposure.
6. Economic implications are significant.
7. Ethical and credibility issues have been prominent and cast in a negative light.
8. Mitigation actions were in progress before the hazard was fully appreciated.
9. The public perception of the danger is a very visceral fear (for example, public schools concerns).
10. Remediation poses its own hazards.
11. Reliability and compensation decisions represent immense fiscal awards.

As the asbestos risk assessment progresses, the associated public health risk estimate has declined (Rowe and Springer 1986). Despite this movement, legislated remediation programs have continued, and extensive investments of public funds have been implemented for removal actions (Mossman et al. 1990). However, quietly and subtly the removal process has emerged as a public health risk in its own right (Abelson 1990). Although public perception is the driving force for action, as with so many of the environmental health risk initiatives, the gap between **perception and fact** is finally becoming public knowledge (Shoedell 1985, Editorial 1986). As the public becomes better informed, perhaps there will be more opportunity for compromise with environmental issues (Ruckleshaus 1983, 1984).

Closing the Gap

Mobilization of local resources and thorough education and communication programs are classic and appropriate solutions to closing "gaps." In SARA Title III [euphemistically called "Community Right to Know"], there is substantive provision of education as an integral element of public planning and response (EPA 1987). The initial step is disclosure of potential environmental hazards by the polluters. This is followed by planning for releases and potential disasters in a public forum with due democratic discourse. Responsible committees with diverse compositions are formed to represent "all" interests, and political compromise leads the planning process in a climate of disclosure and acceptance. Education and openness offer a framework for productive public health action and are a foundation for bridging the ethical gap that so often exists with environmental epidemiologic investigations (Ruckleshaus 1984; Fiore et al. 1990).

THE RISK MANAGEMENT FUNCTION

Risk management is the process of evaluating alternative regulatory actions and selecting among them. Risk management, which is carried out by regulatory agencies under various legislative mandates, is an agency decision-making process that entails consideration of political, social, economic, and engineering information with risk-related information to develop, analyze, and compare regulatory options. The goal of risk management is to select the appropriate regulatory response to a potential chronic health hazard (Figure 10-1). The selection process necessarily requires the use of value judgments on such issues as the acceptability of risk and the reasonableness of the costs to control the exposure (Gordis 1988).

A Real World Example of the Risk Management Process

Basis for Developing a Risk Assessment. The decision to initiate a risk management action with an environmental chemical can be illustrated by the EPA's decision to suspend usage of the preemergent herbicide, 2,4,5-Trichloroethylene (2,4,5-T) in 1979. The EPA is responsible, by statute (the Federal Insecticide, Fungicide and Rodenticide Act (FIFRA)), for regulating pesticides for public safety.

Scientific Evidence to Support a Risk Assessment. In 1977 the EPA began receiving complaints through its Pesticide Incident Monitoring System (PIMS) network from women in an Oregon community suggesting that they were experiencing an unusually high rate of miscarriages. The women were fearful that inadvertent exposure to the herbicide 2,4,5-T was causing these miscarriages. At this time there was also limited information in the scientific literature (McNulty 1977; New Zealand 1977) that suggested a link between 2,4,5-T exposure and spontaneous abortions and spina bifida. Specifically, the experimental literature suggested that if the chemical insult took place within 30 days of conception there was the potential for adverse reproductive effects. Subsequently, an EPA-sponsored health study of miscarriages among this group of women in Oregon also suggested the possibility of a relationship between exposure and miscarriage.

Following a review of these findings, as well as a review of the existing literature, the EPA decided to sponsor a case-comparison study of miscarriages in an area of the United States, where there was public concern about exposure to the chemical. Results from the study indicated that there was a small association ($p < 0.06$) between exposure to 2,4,5-T and an increased rate of miscarriages among the exposed population. The public outcry was loud, and demanding: Environmental groups wanted the chemical banned **immediately**.

Just as vocal, and much more focused, was the position taken by the manufacturer: It was only one epidemiologic study, and it was flawed. The manufacturer took immediate steps to have an expert committee examine the EPA study report. The committee's findings were virtually unanimous: **A substantially flawed study**. Another report by faculty and staff at Oregon State University (partially funded by commercial interests), reviled the study in the strongest terms and found it to be severely flawed. On the other side of the public coin, there were strong opinions voiced by environmentalists that the study was not flawed and should be accepted by the EPA. In fact, this was the only epidemiologic study available for the EPA to use in its risk management of this chemical.

The EPA acknowledged that there were methodological problems with the study:

1. The investigator, independently of the EPA, decided not to conduct the study in the control community, because of the very limited time frame for the study and data analyses.
2. Since there is always great difficulty in ascertaining accurate information on the occurrence of miscarriage in any community, the decision was made to use only women admitted to the hospital for treatment of the miscarriage. This decision was made to ensure that only valid (diagnosed) miscarriages would be entered into the study and that recall bias would not play a part in the results. There was some concern that some of the women of child-bearing age in the exposed area were going to health care facilities away from their community for treatment and delivery (erron-

eously as it turned out), thereby resulting in incomplete ascertainment. There was also concern that since this area was a popular vacation site for many people in that region of the country, pregnant women who were vacationing during the spring and summer months would also increase the numbers of miscarriages although they would not have experienced the requisite exposure. There was also reason to believe that selection bias would skew the analysis since all the women in the community did not elect to go to the hospital for treatment of their miscarriages.
3. Abstracting hospital records can pose serious methodological problems if the persons abstracting the records are not properly trained.
4. Exposure assessment during early stages of the study was particularly weak. In fact, the miscarriage data were generated without an adequate basis for exposure.

The risk management decision by the EPA involved all aspects of the decision-making process. Public opinion was mobilized through the media (TV news, newspapers), headlining the potential hazard to pregnant women exposed to this chemical. Much of the scientific community was clearly at odds on the issue, and there was a dearth of human data to support either position.

Risk Management Strategy. The final issue facing the EPA was the appropriate risk management decision to make. There were clearly scientific concerns over the study design, including the strength and consistency of the epidemiologic data. There was also considerable public opinion, primarily voicing support for stopping the chemical's use.

The EPA, as in all decisions concerning the banning or suspension of an agent must first take into consideration a **risk-benefit analysis**. The more severe the risk, the stronger are the management practices to be employed. On the benefits side, the EPA seeks to determine if there are alternative pesticides available, and if the alternative would be as effective, at the same or reduced costs. Finally, the EPA had to consider whether there were engineering procedures to control the risk, thereby permitting safe use, or whether the benefits were worth the risk.

In the risk management decision on 2,4,5-T the EPA decided that the human risk was substantial (although the study was flawed) and that the consistency of the findings in relation to the animal data supported conservative risk management. The EPA found that an alternative pesticide (2,4-D) existed for satisfactory substitution.

In 1978, on the basis of this less than optimal epidemiologic study, and several animal studies that tended toward similar findings, the EPA decided to **suspend** the use of 2,4,5-T on forest lands in the United States. The decision to suspend was challenged in federal court by the manufacturer, but was later upheld. Subsequently, the manufacturer decided to forego entirely the manufacture of the chemical for sale in the United States.

SUMMARY

In the 2,4,5-T example, communication with the concerned public regarding potential adverse health effects was a focal point of consideration. However, although public opinion is a very potent force, risk management decisions must be made on the basis of the best available scientific evidence. Unfortunately, risk management decisions will seldom be made, totally, on the merits of scientific data. Political, economic, and emotional consideration will always play a significant role in any risk management decision.

Communication with the public, however, is extremely important. The public must be made aware of public policy, and the decision to manage, control, or eliminate a hazardous exposure should always be open to public scrutiny.

ASSIGNMENTS

1. Watch television for "sound bite" approach to environmental health issues.
2. Look in your local newspaper for a human interest story associated with an environmental health problem. See which issues, economic or personal, seem to get the most attention.
3. Define risk management. List and discuss the various components.
4. Think about what you would do to improve the risk management process, vis-a-vis public opinion and scientific issues.
5. Who do you think should be represented on a SARA Title III Community Action Committee formed when there is news of a hazardous waste facility being sited in the area? Provide a rationale for each person or area of expertise to be included.
6. Role play:
   You're the mayor of a small town. You're informed by your congressman that a hazardous waste treatment facility is to be sited in the state, possibly in your community. What will you do? Draw up a plan of action and provide a rationale for each step.
   You're president of a local Parent Teachers Organization (for the elementary school that you attended as a child) when asbestos is found in the insulation around the schools ventilation system. Your town is economically depressed and a major development plan was directed to new floors and desks for the students. Discuss the trade-offs of expending the equivalent of two years' of the schools' budget for removing the asbestos versus

leaving it in place for ten more years when the school building will have been in use for fifty years.

SUGGESTED READINGS

Black, B. 1990. Matching Evidence about Clustered Health Events with Tort Law Requirements. *American Journal of Epidemiology* 132 Suppl (1)S79–S86.

Christoffel, T. & Teret, S. P. 1991. Epidemiology and the Law: Courts and Confidence Intervals. *American Journal of Public Health* 81(12) 1661–66.

Ginzberg, H. M. 1986. Use and Misuse of Epidemiologic Data in the Courtroom: Defining the Limits of Inferential and Particularistic Evidence in Mass Tort Litigation. *American Journal of Law and Medicine* 12(3 & 4): 423–39.

Hoffman, R. E. 1984. The Use of Epidemiologic Data in the Courts. *American Journal of Epidemiology* 123(6): 190–203.

Lilienfeld, D. E. & Black, B. 1986. The Epidemiologist in Court: Some Comments. *American Journal of Epidemiology* 123(6): 961–64.

Lipton, J. P., O'Connor, M., & Sales, B. D. 1991. Rethinking the Admissability of Medical Treatises as Evidence. *American Journal of Law and Medicine* 17(3):209–48.

Rosenberg, D. 1984. The Casual Connection in Mass Exposure Cases: A 'Public Law' Vision of the Tort System. *Harvard Law Review* 97(4):849–929.

Schwartzbauer, E. J. & Shindell, S. 1990. Cancer and the Adjudicative Process: The Interface of Environmental Protection and Toxic Tort Law. *American Journal of Law and Medicine* 14(1): 1–67.

REFERENCES

Abelson, P. H. 1990. "The Asbestos Removal Fiasco." *Science* (247):1017.

Aldrich, T., and Griffith, J. 1991. [Letter to the Editor] RE: Editorial Commentary *American Journal of Epidemiology* 133:512–14.

Brown, P. 1987. "Popular Epidemiology: Community Response to Toxic Waste-Induced Disease in Woburn, Massachusetts." *Science, Technology, & Human Values* 78–85, Summer/Fall.

Brown, P., and Mikkelsen, E. 1990. *No Safe Place.* Berkeley, CA: University of California Press.

Brownlea, A. 1981. "From Public Health to Political Epidemiology." *Social Science and Medicine,* 15D:57–67.

Buffler, P. A. 1988. Epidemiology Needs and Perspectives in Environmental Epidemiology. *Archives of Environmental Epidemiology* 43(2):130–32.

Bullard, R. 1990. "Dumping in Black and White." In: *We Speak for Ourselves: Social Justice, Race and Environment,* ed. D. A. Alston. The Panos Institute, ISBN 1-879358-01-8:4–7.

Cad, R. 1986. "Failing Health and New Prescriptions: Community-Based Approaches to Environmental Risks." In: *Current Health Policy Issues and Alternatives: An Applied Social Science Perspective,* ed by Carol Hill. Athens, Georgia: University of Georgia Press. pp. 53–70.

Conservation Council of North Carolina. 1988. "U.S. Senate Committee Calls EPA Administrator to Task: EPA Action Against North Carolina Call Violation of Law." Chapel Hill.

Cox, L. A., and Ricci, P. F. 1989. "Epidemiology in Environmental Risk Assessment." In: *The Risk Assessment of Environmental and Human Health Hazards: A Textbook of Case Studies,* ed. D. J. Paustenbach. New York; Wiley and Sons: 157–73.

Chase, G. R., Kotin, P., Crump, K., and Mitchell, R. S. 1985. "Evaluation for Compensation of Asbestos-Exposed Individuals II. Apportionment of Risk for Lung Cancer and Mesothelioma." *Journal of Occupational Medicine,* 27(3):189–98.

Editorial, 1986. "The High Risk of Living." *Science* 86:70–71.

Epidemiology Work Group of the Interagency Regulatory Liaison Group. 1981. "Guidelines for Epidemiologic Studies." *American Journal of Epidemiology* 114(5):609–13.

Gordis, L. 1988. *Epidemiology and Health Risk Assessment.* New York: Oxford University Press.

EPA, U.S. Environmental Protection Agency. 1987. *Title III Fact Sheet,* U.S. Government Printing Office 718/810-1302/1280 (1987).

Fiore, B., Hanrahan, L., and Anderson, H. 1990. "State Health Department Response to Disease Cluster Reports: A Protocol for Investigation." *Am J Epidemiol* 132(1) Supplement:S14–S22.

Freudenberg, N. 1984. "Citizen Action for Environmental Health: Report on a Survey of Community Organizations." *American Journal of Public Health* 74: 444–48.

Freundenburg, W. R. 1987. "Perceived Risk, Real Risk: Social Science and the Art of Probabilistic Risk Assessment." *Science* (242):44–49.

Frumkin, H., and Kantrowitz, W. 1987. "Cancer Clusters in the Workplace: An Approach to Investigation." *Journal of Occupational Medicine* 29:949–52.

Glass, R. I. 1986. "New Prospects for Epidemiologic Investigations." *Science* 234:951–56.

Gordis, L. 1988. *Epidemiology and Health Risk Assessment.* New York: Oxford University Press.

Goldsmith, J. R. 1988. "Keynote Address: Improving the Prospects for Environmental Epidemiology." *Archives of Environmental Health* 43(2):69–74.

Gough, M. 1987. "Environmental Epidemiology: Separating Politics and Science." In: *Issues in Science and Technology*: 20–31.

Harris, D. 1984. "Health Department: Enemy or Champion of the People." *American Journal of Public Health* 74:428–30.

Heath, C. W. 1983. "Field Epidemiologic Studies of Populations Exposed to Waste Dumps." *Environmental Health Perspectives* (48):3–7.

Holstein, E., et al. 1984. "Port Allegheny Asbestos Health Program: A Community Response to a Public Health Problem." *Public Health Reports* 99:193–99.

Krauss, C. 1989. "Community Struggles and the Shaping of Democratic Consensus." *Sociological Forum* 4:227–39.

Leitko, T. 1985. "Intergroup Relations in Applied Research: Respondent Participation as a Clinical Intervention." *Clinical Sociology Review* 3:59–71.

Levine, R., and Chitwood, D. D. 1985. "Public Health Investigations of Hazardous

Organic Chemical Waste Disposal in the United States." *Environmental Health Perspectives* (62):415–22.
Lloyd, W. 1983. "Mortality From Brain Tumors and Other Causes in a Cohort of Petrochemical Workers." *Journal of the National Cancer Institute* 70(1):75–81.
Lynn, F. M. 1986. "The Interplay of Science and Values in Assessing and Regulating Environmental Risks." *Science, Technology and Human Values.* 11(2):40–50.
Lynn, F. M. 1987. "Citizen Involvement in Hazardous Waste Sites: Two North Carolina Success Stories." *Environmental Impact Assessment Review.* 7:347–61.
Lynn, F. M. 1990. "Public Participation in Risk Management Decisions: The Right to Define, the Right to Know, and the Right to Act." *Issues in Health and Safety* Spring: 95–101.
Maret, L., and Aldrich, T. "A Compromise Option for Studies of Disease Cluster Reports." *American Journal of Public Health* (In Press).
Masterson-Allen, S., and Brown, P. 1990. "Public Reaction to Toxic Waste Contamination: Analysis of a Social Movement." *International Journal of Health Services* 20:485–500.
McNulty, W. P. 1977. "Toxicity of 2,3,7,8-Tetrachloro-dibenzo-p-dioxin for Rhesus Monkeys: Brief Report." *Bull. Env. Contam. Toxicol.* 18:108–9.
Mossman, B. T., Bignon, J., Corn M., Seaton A., and Gee, J. L. B. 1990. "Asbestos: Scientific Developments and Implications for Public Policy." *Science* (247):294–301.
New Zealand Department of Health. *2,4,5-T and Human Birth Defects.* June 1977.
Nuetra, R., Lipscomb, J., Satin, K., and Shusterman, D. 1991. "Hypotheses to Explain the Higher Symptom Rates Observed Around Hazardous Waste Sites." *Environmental Health Perspectives* 94:31–38.
Ozonoff, D., and Boden, L. 1987. "Truth and Consequences: Health Agency Responses to Environmental Health Problems." *Science, Technology, & Human Values* 70-77, Summer/Fall.
Pretto, E. A., and, Safar, P. 1991. "National Medical Response to Mass Disasters in the United States." *Journal of the American Medical Association* 266(9):1259–62.
Redmond, C. K. 1981. "Sensitive Population Subsets in Relation to Effects of Low Doses." *Environmental Health Perspectives* 42:137–40.
Reich, M. 1983. "Environmental Politics and Science: The Case of PBB Contamination in Michigan." *American Journal of Public Health* 73:302–13.
Rowe, J. N., and Springer, J. A. 1986. "Asbestos Lung Cancer Risk: Comparison of Animal and Human Extrapolation." *Risk Analysis* 6(2):171–80.
Roht, L. H., Vernon, S. R., Weir, F. W. et al. 1985. "Community Exposure to Hazardous Waste Disposal Sites: Assessing Reporting Bias." *American Journal of Epidemiology* 122(3):418–33.
Ruckelshaus, W. 1983. "Science, Risk and Public Policy." *Science* (221) 1026–28.
Ruckelshaus, W. 1984. "Risk in A Free Society." *Risk Analysis* 4(3):157–62.
Schechter, M. T., Spitzer, W. O., Hutcheon, M. E. et al. 1989. "Cancer Downwind from Sour Gas Refineries: The Perception and the Reality of an Epidemic." *Environmental Health Perspectives* (79):283–90.
Shinn, R. 1972. "Science and Ethical Decision." In: *Earth Might Be Far,* ed., I. Barbour, New Jersey: Prentice Hall.

Shoedell, M. 1985. "Risky Business." *Science* 85 43–47.

Shusterman, D., Lipscomb, J., Nuetra, R., and Satin K. 1991. "Symptom Prevalence and Odor-Worry Interaction Near Hazardous Waste Sites." *Environmental Health Perspectives* 94:25–30.

Smith, A. H. 1988. "Epidemiologic Input to Environmental Risk Assessment." *Archives of Environmental Health* 43(2):124–27.

Warner, S. S., and Aldrich, T. E. 1988. "The Status of Cancer Cluster Investigations Undertaken by State Health Departments and the Development of a Standard Approach." *Journal of the American Public Health Association* 78(3):306–7.

Waxweiler, R. J., Alexander, V., Leffingwell, S. S., Haring, M., and Werner, S., and Aldrich, T. 1988. "The Status of Cancer Cluster Investigations Undertaken by State Health Departments and the Development of a Standard Approach." *American Journal of Public Health* 78:306–7.

Wolff, S. 1983. "Problems and Prospects in the Utilization of Cytogenetics to Estimate Exposure at Toxic Chemical Waste Dumps." *Environmental Health Perspectives.* 48:25–27.

Wolman, A. 1986. "Is There a Public Health Function?" *Annual Reviews of Public Health* 7:1–12.

12

Legal Aspects of Environmental Epidemiology

Darlene Meservy, Dr. PH, RN
and Jay A. Meservy, JD

This chapter will discuss:

1. Tort Law as it relates to environmental epidemiology.
2. The role and actions of an epidemiologist in the courtroom.

One outcome from the public's concerns for health risk related to environmental exposures is the potential of eventual litigation. Many of the examples of environmental "incidents" cited in Chapter 1 led to highly visible legal actions. The subject of legal issues is quite subtle and earns its own separate description and discussion.

Environment and the Law

Various government agencies adopt policies related to implementing innovations in society and for determining acceptable risk for those who work or reside in the environment. Many factors influence the technical and political dimensions in assessing and interpreting the acceptability of a given risk in society. Legal recourse is prominent among these factors. To understand the application of environmental epidemiology to litigation one must have a basic grasp of law and how it operates in the United States.

Law is not simply a compilation of legislative inactments but is much broader. In addition to state and federal statutes, the law is recorded in the form of millions of court decisions, relied upon as precedents and printed in hundreds of thousands of bound volumes which in turn are annotated (abstracted by subject matter and factual similarities) in numerous legal research encyclopedias.

Law is divided into civil and criminal categories.

Civil law deals with relationships between individuals and encompasses all matters not specifically designated as crimes. (Entities, such as corporations, partnerships, associations or governmental subdivisions are treated as "individuals" by the law.)

Civil law is divided into two basic categories;
"*codified*"—statutory laws passed by legislative bodies and regulations promulgated by administrative bodies.
"*common law*"—precedents established by prior court decisions that could date back as far as English law.

Criminal law is basically involved with compliance with statutes: that is, was a prohibited action committed? or was a required action performed?.

Each state has a separate jurisdiction, or area of judicial authority, and each has separate and distinct laws, as does the federal government. Although there is substantial "borrowing" of precedents between the various state and federal jurisdictions, there are still major differences between laws of the various jurisdictions. Therefore a separate field of law, the law of **conflicts**, has evolved to set rules for resolving differences between the laws of separate jurisdictions when they clash.

A major field of law within which epidemiology has become an important tool, is the law of **Torts**. The term was derived from the Latin word, *tortious* meaning "twisted," and was, at one time, in common use in English as a synonym for "wrong" (Keeton et al. 1984). A definition of a tort is difficult to find, but Prosser and Keeton offer this one:

> "...a civil wrong, other than breach of contract, for which the court will provide a remedy in the form of an action for damages." (Keeton et al. 1984).

It has been noted that the majority of cancer cases are a result of exposure to man-made chemicals in the environment (Barth and Hunt 1981). The majority of all cancer cases are thus potentially compensable as torts.

Tort law is based on the duty owed by one individual or group of individuals to another individual or group in society. Decisions under tort law use the standard of the "reasonable individual." Nonetheless, as with epidemiologic studies, the proof of cause and effect has proven to be a difficult burden in most instances (for example, courts have been reluctant to hold tobacco companies liable for deaths "known" to have resulted from smoking).

Although the sovereign could not be sued for a tort historically, the shield of **governmental immunity** has, in some instances, been lifted by statute.

These statutes have created carefully defined criteria by which the protection of **governmental immunity** has been waived. (FTCA, 1982.)

Each state and the federal government have two or more levels of courts of original trial jurisdiction. Also within each state and the federal government there are one or more levels of courts with appellate jurisdiction. Appelate courts review cases for accuracy in application of the law to the facts determined during trials in the courts of original jurisdiction or hearings before administrative agencies. They may also, at times, review facts but are generally restricted to correction of abuses of discretion on the part of the trial forum, or to determinations that there was not sufficient evidence to support a particular finding of fact as opposed to judging the weight of conflicting evidence.

"Deciders of fact" (those who make the determination of what the facts are in the given case) may be:

- A jury of one's peers,
- A judge, or
- An administrative law judge (when the matter originates in an administrative agency).

Adjudicators or finders of fact, whether judges, juries, or administrative law judges, are seldom trained in the fields of medicine, epidemiology, and/or the sciences in general.

Judges in the court system, or administrative law judges, may approach a case with no experience in the specific area of law, nor within the area of fact or scientific discipline in which the issues are raised and the facts questioned. As an example, consider the circumstances when the United States Supreme Court reviewed the lowering of the threshold limit value (TLV) for benzene from 10 ppm to 1 ppm by the Occupational Safety and Health Administration (OSHA) (Mortensen and Anderson 1982). The Court was so taxed by the difficulty of "decoding" Congress' approach to the pervasive scientific uncertainty surrounding the risk of benzene that the justices were unable to render a unified majority opinion. Even though five justices did concur that the 1 ppm benzene standard was invalid, they did so in four different opinions. The four minority justices provided a fifth opinion. The five opinions, totaling 120 pages, raised as many questions/concerns as they answered (Mortensen and Anderson, 1982).

In this instance some of the U.S. Supreme Court justices evidenced a lack of the necessary scientific expertise to understand such substantive issues as the irreversible nature of the health effects associated with benzene exposure and the importance of latency in the occurrence of health effects. This lack of comprehension was illustrated by one of the concurring opinions suggesting the parties could have compromised and set a "5 ppm exposure level combined with monitoring to determine if a more stringent standard was needed." This proposition completely ignores issues of the dose-response relationship and shifts the burden of uncertainty to the laborers whom the OSHA Act was designed to protect (Linet and Bailey, 1981).

Similarly, legislators are rarely well informed regarding the complex technical issues necessary to make significant regulatory decisions. Those responsible for passing legislation or establishing regulations may receive mixed messages or contradictory input from the "experts" on vitally needed information. Cases tried before a jury add an additional level of "ignorance." Juries become the deciders of fact, based solely on what transpires within a given court room, under the supervision and guidance of a single judge.

These untrained adjudicators (judges, juries, or administrative law judges), are restricted in their considerations to evidence and arguments actually entered and presented within the specific case. Only the law or policies on which the law is premised is carried from case to case. The evidence considered in a specific case is only that evidence presented by the attorneys representing the parties in that case. The facts are limited and/or controlled by a number of other potentially restricting factors (for example, "rules of evidence" and/or the rulings of the judge)

"Acceptable evidence" is defined by state and federal jurisdictions based on rules of evidence designed to provide information most likely to reach an approximation of the actual facts. Evidence is admitted based on whether it is relevant, material to the issue at hand, or will reasonably lead to relevant or material evidence. It is excluded based on premises or assumptions that it is tainted or unreliable to one extent or another.

"Hearsay" (facts repeated by one individual to another) is an example of evidence that is commonly inadmissable (FRE 801 (c)), although there are exceptions. Other evidence may be admissable in limited circumstances. An example here would be opinion evidence, which is restricted except when the witness qualifies as an expert under the rules (FRE 701, 702, 703, and 704).

Epidemiologic evidence may be subject to interpretation. Evidence actually presented to the fact finder is that information the attorneys believe will be most persuasive or that the judge accepts whether it is the best scientific evidence or not. Two cases involving claims for health damages from Bendectin reached opposite results when one judge excluded evidence that was not significant at a 95% confidence level and another accepted the additional research results (Black and Lilienfeld 1984).

The law sets arbitrary statutory limitations on the time frame within which an action can be commenced. These "statutes of limitation" are intended to guard against unjust claims brought after the ability to defend has been eroded by time. The standard periods of limitation run from two to four years in the field of torts. In tort litigation for cancer occurrence, disease latencies of 20 to 40 years would preclude any such case being brought to trial. Legislatures have added provisions that allow the statutes of limitations to run from the time a person knew, or should have known, of his or her disease and the probable causal relationship with an original exposure to the carcinogen. This allows for access to the courts but does not alleviate the concerns of deteriorated evidence.

The standard of proof in civil cases is a "preponderance of the evidence," often explained to a jury simply as a tipping of the scale (that is, anything more than equal). By contrast the scientific expert is asked to express his or her opinions in terms of reasonable scientific probabilities, usually in terms of a probability statement (p value) or an odds ratio, etcetera. Often such evidence has previously been reviewed by peers as published manuscripts in referred journals. Nonetheless scientists conducting similar studies may differ as to the results and/or conclusions from the studies; sometimes epidemiologists may differ on interpretation of some data (Linet and Bailey, 1981). The wider the gap between the opinions of recognized epidemiologists the more difficult the decision-making process is.

If the epidemiologist is to make a meaningful and beneficial contribution to the judicial process, careful attention must be given to the testimony to be given and how it will be elicited.

The most important element in testifying is preparation. Potential expert witnesses should meet with the attorneys handling the matter in advance and as often as necessary to accomplish several objectives:

1. To assure that the expert understands what the attorney needs by way of testimony.
2. To assure that the expert's opinions are, in fact, consistent with the attorneys' theory of the case and can be used to develop that theory.
3. To assure that the expert has fully and completely educated the attorneys on the subject matter concerning which he or she will testify.
4. To rehearse the expert's testimony with the attorneys. This is necessary so that the expert will be familiar, not only with the intent of the questions that will be asked, but with the style and manner in which they will be asked.
5. To modify scientific terminology and minimize epidemiologic jargon so as to facilitate the communication of the salient ideas in terms familiar to the lay jury and/or judge.

The expert must formulate opinions in advance and carefully educate the attorneys as to what testimony will be needed to support those opinions. On the witness stand the expert must adhere to the rehearsed testimony. If new information comes to the expert's knowledge it must be reviewed with the attorneys before testifying. The introduction of new information at the trial and/or hearing may confuse the attorney and possibly open areas to cross examination inconsistent with the attorneys strategy of the case.

A witness is first questioned by the attorney that requested his or her testimony; this is "direct examination." Then opposing attorneys may **cross examine** the witness; these questions are limited to the ideas and issues raised by the direct examination.

The witness must speak clearly and loud enough so that the finders of fact can hear and understand his or her testimony. Before answering, the witness should be sure the attorney has finished asking a question and be certain he or she understands the question. A witness must not interrupt or talk while a question is being asked or during an objection.

The witness should respond to an opposing attorney's questions briefly; yes or no whenever possible. It is better to admit uncertainty or error, if one has been made, than to be defensive. The expert witness should avoid advocacy and serve as a reporter of the scientific facts as he or she understands them. Whenever possible, the expert must consolidate the supporting data into exhibits. As a rule of thumb, the higher up the policy making ladder one goes, the shorter should be the presentation of the results (Ballick 1981). A simple graph has more meaning to a fact finder or policy maker than a regression analysis.

After testifying, the witness is not free to leave the hearing room unless excused by the judge or administrative law judge. If no further testimony is anticipated the witness may be dismissed, in which event he or she is no longer subject to the subpoena.

At the end of the chapter are listed several suggested readings for a more detailed exploration of this fascinating and challenging aspect of practicing environmental epidemiology.

ASSIGNMENTS

Select a topic of environmental concern. Research it in preparation for potential testimony. If possible, talk with an attorney with trial experience in environmental cases. Write a three-page essay explaining the type of epidemiologic study that would be useful in supporting your (simulated) testimony.

SUGGESTED READINGS

Black, B. 1990. "Matching Evidence About Clustered Health Events with Tort Law Requirements." *American Journal of Epidemiology* 132 Suppl. (1): S79–S86.

Christoffel, T., and Teret, S. P. 1991. "Epidemiology and the Law: Courts and Confidence Intervals." *American Journal of Public Health* 81(12):1661–66.

Ginzburg, H. M. 1986. "Use and Misuse of Epidemiologic Data in the Courtroom:

Defining the Limits of Inferential and Particularistic Evidence in Mass Tort Litigation." *American Journal of Law & Medicine* XII(3&4):423–39.

Hoffman, R. E. 1984. "The Use of Epidemiologic Data in the Courts." *American Journal of Epidemiology* 120(2):190–203.

Lilienfeld, D. E. and Black, B. 1986. "The Epidemiologist in Court: Some Comments." *American Journal of Epidemiology* 123(6):961–64.

Lipton, J. P., O'Connor, M., and Sales, B. D. 1991. "Rethinking the Admissibility of Medical Treatises as Evidence." *American Journal of Law & Medicine* XVII(3):209–48.

Rosenberg, D., 1984. "The Casual Connection in Mass Exposure Cases: A 'Public Law' Vision of the Tort System." *Harvard Law Review* 97(4):849–929.

Schwartzbauer, E. J. and Shindell, S., 1990. "Cancer and the Adjudicative Process: The Interface of Environmental Protection and Toxic Tort Law." *American Journal of Law & Medicine* XIV(1): 1–67.

REFERENCES

Ballick I. H. 1981. Lead: "A Case Study in Interagency Policy Making." *Environmental Health Perspectives* 42:73–79.

Barth, P. and Hunt, H. 1980. "Workers' Compensation and Work-Related Illnesses and Diseases" as cited in Tort Action for Cancer: Deterrence, Compensation and Environmental Carcinogenesis, *The Yale Law Journal* 90:840–62, 1981.

Black B. and Lillienfield D. E. 1984. "Epidemiologic Proof in Toxic Tort Litigation." *Fordham Law Review* 52:732–85.

Federal Rules of Evidence, Rule 801 (c).

Federal Rules of Evidence, Rules 701, 702, 703, and 704.

Federal Tort Claims Act 28 U.S.C. §§1346 (b) & 2671-2680 (1982).

Keeton, W. P., Dobbs, D. B., Keeton, R. E., and Owen, D. B. 1984. *Prosser and Keeton on the Law of Torts*, fifth edition. St. Paul, MN: West Publishing Co.

Linet, M. S. and Bailey, P. E. 1981. "Benzene, Leukemia, and the Supreme Court." *J. Public Health Policy*: June, 118–35.

Mortensen, D. G. and Anderson, F. R. 1982. "The Benzene Decision: Legal and Scientific Uncertainties Compounded." *Legal and Ethical Dilemmas of Occupational Health*, Eds. Lee J. S. and Rom W. N. Pub. Ann Arbor Science: 97–109

"Workers' Compensation and Work-Related Illnesses and Diseases." as cited in "Tort Actions for Cancer: Deterrence, Compensation, and Environmental Carcinogenesis." *The Yale Law Journal*, 90:840–62, 1981.

Index

absolute risk, 40, 249
absorption, 116–118, 120, 126, 176
acceptable risk, 249, 259
acetylation, 167
acetylators, 158, 168
   fast, 168
   slow, 168
acetyl coenzyme (cofactor) A, 168
acetylcholine, 144, 201, 204
acetylcholinesterase, 144
acetyltransferase, 167
Achilles heel (of epidemiology), 168
acrylonitrile, 230–234
act of God, 14
an ACTION (statistical level), 101
activation, 166, 167, 176, 187
   genetic, 187
   oncogenic, 166
active ingredient, 108
activity log, 114
adduct
   DNA, 158–159, 163–164, 169
   hemoglobin, 163, 169
   protein, 163–164, 169
adverse reproductive outcome, 195–199
Agent Orange, 5
(AIDS) Acquired Immune Deficiency Syndrome, 20, 136, 184, 235
adjudicator (finders of fact), 261

Alar, 235
ALERT (statistical level), 101
alert clinician, 98
aliquot, 122
allergic mechanisms, 166
allergic reactions, 208
alkylation, 163
alpha error, 49,53
alpha feto-protein, 147
alpha level, 50, 53
altered structure/function, 156
Alsea (Oregon), 196
ambient environment (defined p.106), 106, 121–122, 197, 226
Ames (see assay), 163
amplified (oncogene) - see activation
amenorrhea, 197
amniotic fluid analysis, 148
analytic epidemiology, 28, 31
anatomic site, 118
anaerobic metabolism, 204
animal testing, 199
anorexia, 206
anoxia, 204
   cytotoxic, 204
   ischemic, 204
angiosarcoma, 22
animal parasites, 19
antibodies, 166
antigens, 166
arylamine, 147, 168
arsenic, 22, 141, 192–193, 200
aryl compounds, 154

aryl hydrocarbon hydroxylase (AHH), 158, 167, 168
asbestos, 22
assay , 21, 146, 170, 172
   Ames, 163
   fungal, 163
   in vivo, 171, 213
   in vitro, 213, 215
   L5178Y lymphoma, 163, 170
   mammalian cells, 163
   rodent, 170
   Salmonella, 163, 170
association, 19, 54, 199
   causal, 54
   consistency, 55
   indirect, 55
   strength, 54
   specificity, 54
assumptions (for risk assessment), 223–227
astrocytes, 200
astroglia, 200
atherosclerosis, 153
at-risk group, 86
athlete's foot, 20
attributable risk, 41
ATSDR (Agency for Toxic Substances and Disease Registry), 7, 100, 208
autonomic ganglia, 201, 206

bacteria, 19
band rinses, 116
baseline level, 249

Index

basal cell carcinoma syndrome, 22
Bendictine, 262
benzene, 22, 193
benzidine, 154, 167
benzo(a)pyrene (BaP), 122, 167
Beta (slope in a linear model), 229, 231
beta error (Type II), 51
bias, 38, 43
　confounding, 38, 44
　misclassification , 36, 106, 175
　observation, 44, 45
　selection, 40, 44
bioaccumulate, 144
bioassay (*see* assay)
bioavailability (defined p., 153)
biochemical marker, 120
biomarker (defined p., 153, NRC p., 155; Chapter, 8)
　battery, 169, 170
　body burden, 158, 170
　genetically determined, 167
　of effect (defined p.156), 174
　of exposure (defined p., 155), 166
　of susceptibility (defined p.156), 167, 168
　oncogenic response, 164, 166–167
　organ-level response, 165–166
　preclinical, 165
　suite, 168, 170
biomaterials, 141
biomonitoring (defined p., 153)
biologic plausibility, 55
biologic dose, 143
biological assays (*see* assays), 21
biological effects, 156
biological markers, 9, 120, 121, 145
　Type I, 152, 158–159
　Type II, 152, 159–167
　Type III, 158, 167–168
biological monitoring, 86, 113, 152–153
biological samples, 113

biologically effective dose, 121
biotransformation, 167
birth defect, 73
blind study, 44
blood, 136, 200, 204
blood-brain barrier (BBB), 200
blood components, 136
Bonferroni procedure, 53
"BOTH" movement, 247
Boulding, Kenneth, 246
Boveri, 160
Broad Street, 16
bromacil, 207

cancer, 21, 22, 184, 186–194, 215
　bad, 190–192
　big, 190–192
　bladder, 165, 192
　brain, 193
　breast, 191–192
　childhood, 197–198
　colo-rectal, 192
　liver, 22, 187, 192, 193
　lung, 22, 184, 190, 192
　oral, 192
　ovarian, 193
　pancreas, 193
　prostate, 191–192
　skin, 193
　stomach, 193
　uterine, 187, 192
cancerphobia, 8, 184
carcinogenesis (ity), 186–194, 219, 225, 226, 229
carboxyhemoglobin, 156, 204
carbon monoxide, 204
carcinogen, 98, 164, 194
　complete, 188
　of the week, 8, 184
　necessary, 188
　pro-, 188
　proximal, 188
　sufficient, 188
cardiovascular disease (CVD), 158, 182
carrier proteins, 157
case-control, 32, 36
causal inference (*see* association), 54, 55
case-pair, 72
cellular damage, 201

cellular repair, 187
cellular targets, 157
central limit theorem (*see* distributions), 50
central nervous system, 20, 196, 200–201
centrifugation, 139
CERCLA - Comprehensive Environmental Recovery, Conservation, and Liability Act, 6
cerebral infarction, 183
cerebrospinal fluid, 200
cerebrovascular disease, 22
chain of custody, 148
Chernobyl, 33
chi-square, 50
chlorobenzilate, 219, 222
chlorine (Cl), 122
chlorinated hydrocarbons, 120
chlorophenoxy residues, 120
cholinesterase (ChE), 43, 121, 142, 159, 173, 204–206
　plasma (PChE), 121, 159, 173, 204–206
　red blood cell (RBChE), 121, 159, 173, 204–206
cholera, 14
choroid plexis, 200
chromosome aberrations, 121, 145, 160, 187, 197
　adducts, 121, 158–159
　chromatid-type, 160
　DNA strand breaks, 145, 160, 161, 187
　micronuclei, 121, 161
　mutation (defined p., 161), 121, 160
　numeric, 160
　structural, 160
chrysene, 122
Cheyne-Stokes respirations, 206
clastogenesis (*see* chromosome aberrations), 187–188
　acentric fragments, 187
　breaks, 187
　deletions, 187
　dicentrics, 187
　gaps, 187
　quadriradial, 187
　rings, 187

Index 269

clinical trials, 28
closeness, 67
cluster, 62, 245
  analyses, 70–76, 94–97
    Barton, 72
    Chen, 72, 76
    CuSum, 73, 76
    Grimson, 73
    Knox, 72
    Ohno, 73
    Person, 72
    Poisson, 73, 76
    REMSA, 72, 76, 96–97
    Scan, 72
    Sets, 73
    Texas, 73, 76
  joint investigations, 245
  CLUSTER, 63–73, 85–96
coffee, 22
cohort (defined p.32)
  retrospective, 32
  prospective, 34, 39, 175
  historical prospective, 35, 175
colchicine, 146
coma, 204
complex mixtures, 107, 111, 188–189, 223
community action groups, 243, 245
community intervention trials, 28
"community right to know" (SARA Title III), 251
concentration (*see* exposure intensity), 113, 114, 143, 144
condensable organic material, 122
confidence, 48
  interval, 48, 49
  limit, 49, 94, 230
confounding, 37, 38, 44, 57
congenital anomalies, 198
congenital malformations, 195, 197, 199
conjugation, 187
contact dermatitis, 166
continuum of events, 156, 170
control(s)
  laboratory, 147
  negative, 133
  on-site, 147
  positive, 133
corium, 117

coronary artery disease, 21
cramps, 207
creatinine, 139
credibility gap, 249
cross-contamination, 140
cross-examination (*see* examination), 264
cross-sectional, 32, 38
cyanide, 204
cyanosis, 206
cytochrome P450 (*see* P450 group), 159, 167

data sources, 84, 90
  primary, 90
  secondary, 90
deciders of fact, 261
Delaney clause, 226
Density Equalized Map Projection (DEMP), 96
dependent variable, 118, 171
derived standard, 133
dermal exposure, 113, 115, 117, 221
dermatitis, 165
descriptive epidemiology, 28, 29
determinism, 18
detoxification pathway, 225
diazinon, 207
differentiation (cellular), 196
dioxin, 3, 22, 216
diphtheria, 21
diplopia, 207
disease, 19, 29
distribution, 15, 50, 176
  binomial, 50
  chi-square, 50
  log-normal, 50
  normal, 50, 176
  normally, 50, 51
diuron, 207
DBCP (dibromochloropropane), 39, 198
DEDTP (diethyl dithiophosphate), 31
DEP (diethyl phosphate), 31, 38, 39
DETP (diethyl thiophosphate), 31
DMP (dimethyl phosphate), 31
DMTP (dimethyl thiophosphate), 31
DMDTP (dimethyl dithiophosphate), 31
DNA strand breaks, 121, 145, 160, 161
dose, 118, 126, 133, 143, 224
  absorbed, 116–118, 156
  biological effective, 121, 143, 156
  cumulative, 224
  effective, 156
  integrated, 224
  target, 156, 164
dose-response, 9, 55, 118–120, 126, 216, 218, 224, 226, 231
  relationship (defined p., 118)
  estimation, 215, 231
dosimetry, 155
double blind study, 44
Drosophila (*see* assay), 163
dry ice, 139
dyspnea, 206

early response (biomarker), 156
ecologic (study), 40
ecologic fallacy, 40, 92
  (example on p., 92)
ecosystem (*see* unified ecosystem), 247
eczematous reaction, 166
effect modifiers, 157
effective dose (50%)(ED50), 225
egalitarianism, 245
electrolytes, 200
electrophile, 164
electromagnetic fields, 22
embryo, 195
embryogenesis, 197
embryotoxic, 199
endothelial cells, 201
endpoint, 57
environment, 14, 106, 195, 204
  ambient, 106, 121, 197, 226
  micro, 114, 121, 127
environmental
  disasters, 1, 196, 208
  epidemiology (defined p.22–23), 168, 186
  equity, 247
  ethics, 247
  inequity, 247
  pathology, 147
environmentalists, 195, 252

enzyme-linked
  immunosorbent assays
  (ELISA), 164
epidemiology, 18, 22, 167, 199
  analytic, 28, 31
  biochemical, 152
  descriptive, 28, 29
  experimental, 28, 50
  molecular, 152
  observational, 28, 50
  reasoning, 55
  shoe-leather, 248
epidermal, 117
epoxides (hyperactive), 168, 188
error, 43, 50, 51
  alpha, 49, 53
  beta, 51
  observer, 45
  measurement, 43
  method, 45
  random, 43
  standard, 50, 51
  systematic, 43
  type I, 49, 51
  type II, 51
erythema, 165
erythrocytes, 142, 163
ethylene oxide, 142, 163
etiology, 190, 192, 197
evidence, 262
  acceptable, 262
  epidemiologic, 262
  exhibits (data), 264
  inadmissible, 262
  opinion, 262
  preponderance of, 263
  rules of, 262
examination (legal), 264
  direct, 264
  cross, 264
exhibits (legal), 264
exogenous (agents), 196, 200
experimental epidemiology, 28, 50
expert (testimony), 262
exponential transformation, 51
exposure, 29, (Chapter, 6), 114, 126, 168–170, 174, 195, 213, 223
  chacterization, 105
  dermal, 113, 115, 117, 221
  duration, 105, 216
  high end, 126

intensity, 113, 216
magnitude, 105, 213
oral, 113, 116
pattern, 216
respiratory, 113, 208
route, 105, 108, 127, 188, 208
schedule, 216
scenario, 126
exposure assessment, 9, 105, 168, 215–217
extraction method, 143
extracellular fluid, 200
extrapolation, 213, 223, 225, 227, 228, 230

false negative, 46
  in a risk assessment, 225
false positive, 46, 100
  in a risk assessment, 225
fasciculation, 207
fatty acids, 22
Farr, William, 15
Federal Insecticide, Fungicide and Rodenticide Act (FIFRA), 251
fetal calf serum, 146
fetus, 196
field trials, 28
finders of fact, 261
fine particle, 122, 125
5-bromodeoxyuridine (BrdUrd), 146
fixed location monitor, 126
follicle stimulating hormone (FSH), 148
formalin, 136
4-aminobiphenyl, 167
Fracastorius, Hieronymus, 15
fragile sites, 186
fungal assays (*see* assays), 163

Gas Chromatography/Mass Spectrometer (GC/MS), 122
genetic markers, 145
  cellular, 145
  molecular, 145
genetic regulation, 187
geographic cells, 72
geo-cell, 72, 78
generic human, 224
genome, 186
genotoxicity, 145

germ theory, 15
Giemsa stain, 147
glia, 200–201
glial cells, 200
glucose-6-phosphatase dehydrogenase (G6PD), 147
glutathione-S-transferase (GST), 158, 168, 187–188
goodness-of-fit (test), 51, 78
governmental immunity, 260
Graunt, John, 15

half-life, 224
hazard identification, 9, 170, 215–217, 250
hearsay (evidence), 262
helium, 123
HeLa (*see* assay), 163
Henle, Jacob, 15
hepatitis, 136
high end exposure measurement, 126
high pressure liquid chromatography (HPLC), 143
Hippocrates, 14
HIV (human immuno virus), preventing transmission, 149
human interest, 242
human subjects review board, 133
hyperhidrosis, 207
hyperplasia, 158, 171
hypertension, 158
hypoxia, 201
hypothalamic, 200
hypothalamus, 200–201

Ice, 137, 139
Incidents Environmental, 2, 196
  Three mile Island, 2, 196
  Love Canal, 2, 196
  Chernobyl, 3
  Seveso, 3, 196
  Times Beach, 4
  Woburn, 5
inconclusive study (*see* statistical power), 51
independent variable, 118, 171
index case, 99
induction, 194

indirect association, 55
infection, 19, 190
influenza, 15, 20
informed consent, 133
initiation, 186, 188, 224
integument (*see* dermal absorption), 117, 165
interaction, 132, 188
interactive products, 159
intercept, 96
intercept point, 227, 228
interstitial fluid, 200
intoxicants, 204
intracell, 96
intraindividual differences, 164
in vivo, 171, 213
in vitro, 213, 215
involuntary risk, 186
ionizing radiation, 188, 192–193, 197
irreversible effects, 201, 208, 261
ischemic heart disease, 182

joint investigation, 245
jurisdiction, 260, 261

Koch, Robert, 18, 54

laboratory practice Chapter, 7, 148
lacrimation, 207
latency, 21, 175, 194, 261
law Chapter, 12
  civil, 260
  codified, 260
  common, 260
  conflicts, 260
  criminal, 260
  tort, 259–260
lead, 116, 158, 196, 198, 217, 224
lepers, 15
lethal dose (50%) (LD50), 215, 225
leukemia, 22, 193
Limit of Detection (LOD), 127
limitation, 262
  statutory, 262
  statutes of, 262
lipid, 118
Lister, Joseph, 17
litigation, 259
local organizations, 242

logarithmic scale, 51
longitudinal, 32
Love Canal, 2, 196
luteinizing hormone (LH), 148
lymphoma, 170, 193
lymphocytes (*see* White Blood Cells), 163

macromolecule, 136
malaria, 14
malononitrile, 204
mammalian cell cultural assays (*see* assays), 163
markers (*see* biological markers, biochemical markers, bimarkers, genetic markers)
matched (cases and controls), 37
Maximum Contamination Level (MCL), 5
Maximum Daily Intake (MDI), 235
Maximum Individual Risk (MIR), 235
measles, 20
measurement error, 43
Media, 241–42
  influence, 8, 184
medical surveillance, 100
membrane permeability, 157
mercury, 196, 217
mesothelioma, 22, 250
metabolism, 159, 168, 176, 187, 204, 227
  Phase I (early stage), 159, 167, 168
  Phase II (late stage), 168
metastases, 166, 186
methemoglobin, 204
Method of Detection limit (MOD), 127
methyl mercury, 196, 217
methylene, 122
miasma, 14
microenvironment, 114, 121, 127
micronuclei, 121, 161
migrants (cancer among), 191
Mill's Canons, 19
Minamata Bay, 196
miosis, 206
miscarriage, 251
misclassification, 36, 100, 175

mitigation, 247
Mixed Function Oxidase (MFO) (*see* P450 Group), 159
models, 164, 170, 215, 224, 227–234
  curvilinear, 228
  extrapolation, 228
  linear, 224, 228, 233
  mathematical, 227
  Mantel-Bryan log-probit, 227
  multi-hit, 227
  nonlinear, 228
  single hit (or one hit), 227, 228
monitoring, 113–116, 121–124, 221
  area, 114
  fixed location, 126
  personal, 114, 221
moving window, 72
multisite studies, 111
mumps, 20
mutagens, 187
mutations, 121, 145, 159, 161, 197
  base pair substitutions, 163
  frameshift, 163
  point, 145, 163
myelinated axons, 201
myocardial infarction, 183
myopathy, 207

N-acetyltransferase (Nat), 147, 154, 167–168
naphthalene, 122
  perdeuterated, 122
National Ambient Air Quality Standard (NAAQS), 111, 122
National Priority Site (NPL) - *see* SUPERFUND, 108
National Research Council (NRC), 155
natural experiment, 28
nearest neighbor, 96
negative control, 133
negative study (*see* statistical power), 51
nested study, 34, 57, 100
neuroblastoma, 166
neuromuscular system, 200, 204

Index

neurologic disease, 22, 200–207, 215
neurons, 200–201
neurotoxic, 39, 186, 200–207
neurotoxicants, 200–207
neurotransmission, 144
NHANES (National Health and Nutrition Examination Survey), 31, 38, 217
nickel/chrome alloy, 135
nitrogen dioxide, 111
nitrites, 204
Nixon, Richard M., 21
nonionizing radiation, 193, 198
nonreversible effects, 208
normal distribution, 50
North Carolina rule, 245–247
nosocomial infections, 84
Not In My Back Yard (NIMBY), 246
nucleolus, 201

observation bias, 44, 45
observational epidemiology, 28, 50
observer error, 45
occupancy, 65
Occupational Safety and Health Act (OSHA), 262
oligodendrocyte, 200
oncogene, 166, 171, 186
    activation, 166
    amplified, 166
    N-myc, 166
    protein products, 166
    proto-, 166
one-hit (see single hit model)
one-in-a-million (1 x, 10-6 or 0.000001), 229, 231
oral exposure, 113, 116
organogenesis, 198
organophosphate (OP), 31, 43, 120, 144, 159, 204–207
organ-level responses, 165
organ dysfunctions, 158, 165
organic material, 122
    condensable, 122
    volatile, 122
    semivolatile, 122
ORM-A, 140
oxidative enzyme systems, 204

ozone, 111

p-value, 57
P450 group (liver enzymes), 145, 159, 167
paradigm, 213
parameters, 228
parathion, 118, 173
parasite, 19
parenchyma, 200
paresthesia, 217
particulate(s), 122
Pasteur, Louis, 18
patch techniques, 115
pathobiologic effects, 133
pathway, 127, 188
Penicillin-Streptomycin solution, 146
perceived problem, 242
perception, 242
    public, 1, 8, 184, 186, 249, 250
    popular, 1
    (intolerance of ambiguity), 243
percutaneous, 117
peripheral neurons, 201
peripheral nervous system, 201
persistence, 155
personal air sampler, 114
person-to-person, 14
person-year, 35, 46, 57
personal interaction, 242
personalize risk, 242
pertussis, 21
Pesticide Incident Monitoring System (PIMS), 251
pesticide poisoning, 108, 207
pharmacodynamic, 164
pharmacokinetics, 164, 176
phenanthrene, 122
phenotypic, 168, 199
phlebotomy, 136
phosphine, 166
phosphorylation, 206
Phytohemaglutinin, 146
plague, 14
point mutations, 145
point source, 208, 221–222
Poisson, 50
poliomyelitis, 20
political gap (in risk management), 249
poly aromatic hydrocarbons (PAHs), 122, 163, 167, 193
polychlorinated byphenyls (PCBs), 145, 158, 198
population, 18, 43, 45, 49, 50
    derived, 50
    reference, 43, 50
    target, 43
positive control, 133
potassium, 146
potassium chloride, 146
power (statistical), 51
precedent (legal), 259
precision, 43
precision of measurement, 43
preparation of [laboratory] materials, 135
prevention, 8, 9, 171, 197
preventive measures, 183
probability, 230
probability cell, 72, 78
probabilistic, 19, 236
procarcinogen, 188
promotion, 186, 188
proof (see evidnece)
proportional incidence ratio (PIR), 47
proportional mortality ratio (PMR), 47
proximal carcinogen, 188
public, 8, 240, 244
    policy, 8, 254
    scrutiny, 243, 254

quality assurance, 148
quality control (laboratory), 141
quarantine, 20

rabies, 20
radiation, 135, 188, 192–193, 197, 198
    ionizing, 188, 192–193, 197, 198
    nonionizing, 193, 198
radon (gas), 22, 193
random error, 43
random variation, 53
regulation, 187
    cellular, 187
    genetic, 187
reaction, 132
reasonable
    individual, 260
    person, 244

Index

reasoning (epidemiologic), 55
Rebuttable Presumption Against Registration (RPAR), 219–220
Red Blood Cells (see erythrocytes)
regulatory process, 215
relative risk, 40, 51
reliability (see error), 43, 45
remedial action, 7
remediation, 197
REMSA, 72, 76, 96–97
repair enzymes, 157
repair (cellular), 187
reproductive toxins, 195–199, 215
residue analysis, 155
Resource Conservation and Recovery Act (RCRA), 246
respired air, 142
respirable suspended particulates (RSP), 111
respiratory effects, 208
respiratory exposure, 113, 208
restricted use, 220
reversible effects, 208
rheumatic fever, 21
Rickettsiae, 20
risk, 57, Chapter, 10, 212–237
  absolute, 40, 249
  acceptable, 249, 259
  assessment, 7, 9, 36, 102, Chapter, 10, 219 (defined p., 213)
  attributable, 41, 43, 194
  characterization, 9, 215, 217–223
  communication, 241
  involuntary, 186
  management, 223, 251 (defined p, 251)
  relative, 40
  voluntary, 184
  zero, 249, 259
risk-benefit analysis, 253
Rocky Mountain Spotted Fever, 20
rough endoplasmic reticulum (RER), 201
route of exposure, 105, 108, 127, 188, 208
rubella, 195, 198

saccharin, 22
Salk vaccine, 20
Salmonella assay (see assay), 163, 170
salivation, 207
sample, 43, 50
sample size, 51
SARA (Superfund Amendments and Reauthorization Act), 6, 251
SARA Title III ("Community-Right-to-Know"), 251
scarlet fever, 15
Schwann cells, 200
Science Advisory Panel (SAP), 220
scientific notation, 230
scientific measurement, 249
screening, 184
  cluster reports, 66
segmental analysis, 141
selection, 40, 44
sentinel events, 6, 76, 98
serotonin, 201
Seveso, 3, 196
seminal fluid, 142
semivolatile organic material, 122
sensitivity, 45
  exposure characterization, 107
  of biomarker, 163
  of laboratory assay, 142
shelf-life, 136
significance, 48, 52, 244
  practical, 243
  public health, 244
  statistical, 29, 48, 50–53, 58, 101, 182, 243–244
  tests, 48, 52
single cell gel (SCG), 161
Sister Chromatid Exchange (SCE), 121, 145, 160, 171
site profile, 221
smallpox, 14, 20, 21
smoking, 22, 183–184, 192–193 (see lung cancer)
Snow, John, 15
social accountability, 247
somatic cell mutation, 160
spaceship (Earth), 246
Spatial Auto-Correlation (SAC), 95
spatial monitors, 114
specificity, 45
  exposure characterization, 107
  of biomarker, 163, 172–174
  of laboratory assay, 142
specimen collection, 135
spectrum of health effects, 208
sperm, 121, 142, 197
  collection, 142
  counts, 147
  density, 148
  morphology, 121, 147
  motility, 148
  viability, 147
spiked sample, 141
spontaneous abortions (SA), 47, 197
square root transformation, 51
standard man, 224
surface area, 118
standardized mortality ratio (SMR), 46, 51, 101
standardized incident ratio (SIR), 46, 48
standard error, 50, 51
statistical tests (examples), 50, 58
statutory limits (see limitations)
strength of association (see statistical significance)
studies (see also epidemiology and specific study designs, Chapter, 3)
  analytic, 28, 31
  blinded (single, double), 44
  cohort, 32, 34–35, 39
  case-control, 32, 36
  cross-sectional, 32, 38
  descriptive, 28, 29
  ecologic, 40
  inconclusive, 51
  longitudinal, 32
  nested, 34, 57, 100
  negative, 51
  prospective, 34, 39
  observational (ecologic), 28, 50
  retrospective, 32, 35

subgroup analysis, 62
subpoena, 264
sulphur, 122
sulphur dioxide, 111
Superfund *see* CERCLA - Comprehensive Environmental Recovery, Conservation, and Liability Act, 6, 108
Superfund site, 108
surveillance (defined) p.83, Chapter, 5, 91, 95-97
  active approach, 85
  data (applications) p.86
  elements, 84
  epidemiologic, 84, 100
  medical, 100
  passive approach, 85
  public health, 85, 100
survey analysis, 86
susceptibility, 22, 167, 188, 208
swab (swabbing), 116
Sydenham, Thomas, 15
systematic error (*see* bais), 43
system dysfunction, 157

target population, 43
TCDD (2,3,7,8-tetrachloro dibenzo-para-dioxin), 3, 22, 216
temporal (structure, relationship), 55, 174
teratogen, 98, 198
teratogenic agents, 198, 199
testosterone, 148
tetraethyl lead, 116
tetraplegia, 207
thalidomide, 196
Third World (cancer in the), 191
3-methylcholanthrene, 167

^{32}P-postlabeling, 144, 163
Three Mile Island, 2, 196
threshold, 198, 224, 227, 228
Times Beach, 4
time series, 73, 94
Tort Law (defined p., 259), 260
toxins, 208
  reproductive, 196
toxicant(s), 114, 116, 143, 196, 208
toxicologic, 213
Tradescantia (*see* assays), 163
transport of (laboratory) materials, 135
Trichlorethylene (TCE), 5
trihalomethane (THM), 22
tuberculosis (TB), 20
2-aminoaphthalene, 152
2,4-Dichloroethylene (2,4-D), 253
2,4,5 trichloro phenoxyacetic acid, 120
2,4,5-Trichloroethylene (2,4,5-T), 5, 251, 253
two-step decision process (technique), 73, 101
2,3,7,8-tetrachloro dibenzo-para-dioxin (dioxin), 3, 216
typhus fever, 20
type I error, 49, 51, 244
type II error, 51

ultrasensitive enzymatic radioimmunassays (USERIA), 164
ultraviolet, 135
unequal areas, 96
unified ecosystem, 247
unit risk, 230, 234
urine , 139-140, 169
  collection practices, 139

first voided morning, 139
sequential specimens, 139
urinary metabolites, 169
uterus, 195

vaccination, 20
vacutainer tubes, 137
validity (*see* also bias and error), 43, 45, 148
variables, 85, 118, 191-192
  dependent, 118, 171-172
  independent, 118, 171-172
  primary, 85
  secondary, 85
vascularization, 201
vesicles (*see* dermal absorption), 117, 165-166, 201
vinyl chloride, 22, 187-188
vital statistics, 15
Vitamin A, 192
virus, 19, 188, 193, 195, 198
volatile organic material, 122
voluntary risk, 184
von Plenciz, Anton, 15

War on Cancer, 21
Webster, Noah, 15
wet ice, 137
witness (legal), 264
worse case, 224, 235
white blood cells (WBC), 144 (*see* lymphocytes)
Woburn, 5

xenobiotic, 144, 147, 159, 170

yellow fever, 14
Yugoslavia, 124

Zea (*see* assay), 163
zero risk, 249
zonulae occludents, 201
zymbal gland, 229, 234